SOIL EROSION IN A COASTAL RIVER BASIN

A CASE STUDY FROM THE PHILIPPINES

by

Random DuBois

Development Alternatives, Inc.

UNIVERSITY OF CHICAGO GEOGRAPHY RESEARCH PAPER NO. 232

The University of Chicago
Committee on Geographical Studies
1990

Library of Congress Cataloging-in-Publication Data

DuBois, Random
Soil erosion in a coastal river basin : a case study from the Philippines / by Random DuBois.
p. cm. -- (University of Chicago geography research paper ; no. 232)
Includes bibliographical references (p.) and index.
ISBN 0-89065-139-6 (paperback) :
1. Water--Pollution--Philippines--Siquijor (Province) 2. Land use--Environmental aspects--Philippines Siquijor (Province) I. Title. II. Series.
TD313.P6D83 1990
628.1'683'095995--dc20 90-11221

Geography Research Papers are available from:

The University of Chicago
Committee on Geographical Studies
5828 South University Avenue
Chicago, Illinois, USA 60637-1583

For Sherry, Alejandra,
and Byron

ANDERSONIAN LIBRARY
WITHDRAWN
FROM
LIBRARY
STOCK
UNIVERSITY OF STRATHCLYDE

Contents

Figures

Tables

TABLES

ACKNOWLEDGMENTS

Grateful acknowledgment is made to the Government of the Philippines/World Bank CVRP Project, whose financial support of the research brought the study to fruition. Beyond this, there exist a number of individuals whose support was particularly crucial for the completion of the study. These include: John Dalton and Ed Queblatin for their overall support of the project; Angel Alcala, "Sander" Calibo, June Calinawan, Vic Evalle, Emie Langcoyan, T.F. Luchavez, "Soy" Magsayo, June Merced, Chris Quimbo, Ed Sibonga, Fred Vandavusse, Henny Zerna, "Ting-ting," and Alan, who all contributed in one or more aspects of the field operations; Thomas Dunne, Norton Ginsburg, Marvin Mikesell, Joseph O'Reilly, and James Wescoat, for guidance, both conceptual and technical; and John Clark, Samuel Kunkle, and Edward Towle, who graciously agreed to serve as outside readers. In addition I would like to acknowledge all the other individuals in the CVRP/Siquijor upland and coastal site management units, Siquijor Provincial Government staff, Silliman University Marine Laboratory, University of San Carlos Centers for Population Studies and Water Resources, ACIPHIL, CVRP/Cebu, CERTEZA, the Bureau of Soils, Region 7, and a host of other government agencies who assisted the project in a variety of ways.

Chapter 1

INTRODUCTION

A coast can be defined as an environmental zone characterized by the interface of aerial, terrestrial, and marine biospheres. Generally, these areas are inherently diverse and complex in terms of their community composition, biological productivity, and natural processes. Moreover, they are typically hazardous in nature, subject to physical events originating from both offshore and inland source areas. Finally, many of the world's coasts are coming under increased pressure to support a broad range of human land use (and marine) activities, uses that can come into conflict with one another and with the basic ecological processes of the coast itself.

The coast's inherent complexity and hazardous nature, together with intensified human use, has created a growing need for sound planning. To date, much of the effort in planning and management of coastal areas has been directed toward mitigating existing or potential resource conflicts associated with competing resource uses in the coastal zone proper. In contrast, very little appears to have been done in addressing problems associated with off-site resource utilization. These distant problems can be broadly classified into four types based on origin: (1) offshore/on-shore (e.g., oil spills); (2) longshore (e.g., thermal pollution from energy-generating plants located in the coastal zone); (3) aerial-borne (e.g., acid rain); and (4) upstream/downstream (e.g., degradation of coastal water quality associated with upstream industrial discharges).

Although all four sets of conflicts are believed to be significant in their magnitude and range of effect on coastal areas, it is the last that is viewed as most critical and in need of further study. This conclusion, in part, is due to the growing body of empirical evidence demonstrating the coastal environmental impacts and associated high economic costs attributable to river basin development.[1] Poor planning, in combination with

---

[1] For purposes of the present study the terms "watershed" and "river basin" will be used interchangeably to mean a natural landform interconnected by the unidirectional flow of sur-

the watershed's biophysical interdependencies, bounded nature, and unidirectional flow of water, in effect serves to channel many off-site impacts into the coastal environment. Typical downstream issues associated with upland land uses include soil toxification, land instability and river shallowing, increased risk of natural hazards, saltwater intrusion, and pollution effects.

Moreover, the satisfactory resolution of these upstream/downstream issues is particularly challenging owing to a number of constraints inherent in the conflict. These include impediments to: the timely discernment of downstream effects; the identification of the relevant transport pathways connecting source area with site of impact; the accurate characterization of the underlying socioeconomic factors serving as driving forces behind the physical source(s) of conflict; and the formulation of responsive institutional measures and arrangements needed for issue resolution.

Heretofore, much of the research on upstream-downstream conflicts has been directed toward the more dramatic cause-and-effect impacts associated with large-scale, capital-intensive development projects. Perhaps the best documented examples were the coastal impacts attributed to the upstream entrapment of nutrients and sediment following the construction of Egypt's New Aswan High Dam. These included nutrient depletion, accelerated coastal erosion, and fisheries declines.[2] Moreover, until recently much of the research focus has been dominated by the study of the biophysical characteristics and processes associated with the conflict and only secondarily has addressed the socioeconomic dimension. There is, however, another aspect to the issue that appears to have been neglected, that is, the downstream effects associated with increasing human encroachment and exploitation of upland marginal lands primarily in the forms of slash-and-burn farming, overgrazing, and logging practices, typical of many tropical developing countries.[3] This is particularly incongruous given the enor-

---

face water, which, owing to the combination of gravity and physical boundaries, is transported from the system through one exit.

2 One effect involved a reduction in downstream nutrient levels contributing to a 60 percent decline of the eastern Mediterranean sardine fishery which, previous to dam construction, averaged 18,000 mtons annually. Other impacts attributed to nutrient and sediment reductions included declining soil fertility in the lower banks of the Nile, declines in shrimp production, and the alteration of the coastal sediment budget from one dynamically balanced between forces of deposition and erosion to one dominated by erosion. See H.M. Fahim, *Dams, People and Development* (New York: Pergamon Press, 1984), pp. 37-38; and Carl J. George, "The Role of the Aswan High Dam in Changing the Fisheries of the Southeastern Mediterranean," in *The Careless Technology*, ed. M. Taghi Farver and John P. Milton (Garden City: Natural History Press, 1972), p. 160.

3 D.D. Richter, S.R. Saplaco, and P.F. Nowak, "Watershed Management Problems in Humid Tropical Uplands," *Nature and Resources* 21 (October 1985): 11.

mous amount of research interest concerning the social dynamics behind these practices.

Tropical coastal ecosystems (such as coral reefs, marine grassbeds, mangroves) are some of the most productive communities in the world. Moreover, in many countries these areas are of particular socioeconomic importance as sources of food, livelihood, and a wide range of other uses. However, many of these same communities are characterized by low environmental thresholds to physiochemical perturbations. Inland, poverty, high population growth rates, and intensive exploitation of coastal lowlands are sustaining rapid and typically unplanned development in the interior highlands of many tropical developing countries. The accompanying modifications to land and water quality are increasingly placing downstream ecosystems (and the livelihoods they support) at risk. Moreover, the range, extent, and significance of coastal impacts associated with the exploitation of marginal lands are largely unknown. Earlier studies of the capital-intensive projects appear to bear little relevance to this subset of issues. For example, project-related impacts are associated with site-specific rather than diffuse modifications of the landscape (nonpoint source areas). Second, the underlying social factors contributing to human interventions in upland areas are decidedly different. Finally, the ranges of institutional responses available to mitigate downstream impacts differ significantly.

These two sets of factors characteristic of many tropical coastal watersheds form the basis of the research problem: (1) increasing encroachment of upland marginal lands in the presence of ecologically and economically important coastal ecosystems, and (2) the absence of needed information with regard to range and magnitude of impact and the underlying social-economic processes contributing to the conflict.

To study this aspect of upstream-downstream conflict requires, at minimum, four sets of information: (1) characterization of the zones of impact at both the source and site of impact; (2) identification of the relevant biophysical processes responsible for the off-site impacts on the coast; (3) characterization of the underlying socioeconomic processes responsible for the issue; and (4) definition of the institutional framework within which the issue occurs and must be resolved. Specific topics to be examined in the present study are: (1) the nature, significance, and implications of nonsustainable watershed land-use practices on coastal resource-user systems; (2) the perceptions and knowledge of upland and coastal resource users regarding land-use practices and river basin processes that in combination contribute to conflict; (3) the adjustments employed by coastal resource users in response to changing environmental conditions attributable to upstream land-use practices; (4) social forces contributing to existing land-use practices; and (5) selected policies that may be contributing to or mitigating the conflict.

Owing to the paucity of information and a limited budget, the approach selected for the research was the case study. The location of the study was a small coastal watershed in the Central Visayan island-province of Siquijor in the Philippines. The specific approach entailed two components: (1) the conduct of an environmental assessment and year-long biophysical monitoring program; and (2) the development and use of a survey questionnaire together with a number of special studies designed to examine specific social questions pertaining to land use, environmental perceptions, social dynamics, and policies relevant to intrabasin resource issue conflicts. The information derived from these two components, together with a detailed historical literature review, provided the basis for the subsequent integrative analysis and conclusions.

Chapter 2

CONSIDERATIONS, CONSTRAINTS, AND APPROACHES TO THE INTERDISCIPLINARY INQUIRY OF ENVIRONMENTAL DEGRADATION

Much of the existing body of knowledge concerning environmental degradation in tropical regions comes from impact studies of natural scientists.[1] However, to address fully the complexity of the problem described in chapter 1 requires not only an adequate understanding of the biophysical characteristics and processes involved in the conflict, but also a full and equal treatment of the underlying social processes. Clearly, then, the approach to the study must be interdisciplinary in nature, integrating relevant data from both the social and natural sciences. Chapter 2 represents a compilation of pertinent materials that served as the basis for the study's research design (chapter 4). Its organization has been based broadly on three key constraints facing researchers (and others) in the study of environmental degradation. These were identified in a recent, seminal work on the subject as: (1) fundamental, often ideological, disagreements about the causes and significance of land degradation; (2) an inability to integrate the relevant analytical tools from the natural and social sciences into a framework capable of addressing the issue; and (3) failure to view degradation within a wide historical and geographical context.[2]

Structural Factors

Interest in natural resources utilization and its relationship to environmental quality has changed markedly over the course of time. Based on the written record, Western concern over the despoliation of the environ-

---

1 It has been noted that a number of social processes appear to be responsible for nonsustainable exploitation of the environment but, with the exception of population growth, remain largely unexplored. See Piers Blaikie, *The Political Economy of Soil Erosion in Developing Countries* (New York: Longman Scientific and Technical, 1987), p. 24.

2 Piers Blaikie and Harold Brookfield, introduction to *Land Degradation and Society*, ed. Piers Blaikie and Harold Brookfield (London: Methuen and Co., 1987), pp. xvii-xxi.

ment extends back at least as far as early Greece.[3] By the Middle Ages this concern was manifested in a number of laws regulating the cutting of forests and grazing densities on mountainsides in Central Europe that demonstrated a relatively sophisticated understanding of the relationships between land use and physical processes.[4] In the eighteenth century the works of the French engineer Jean Antoine Fabre provided further evidence about upland-lowland processes and land-use conflicts resulting in excessive runoff and downstream flooding.[5]

In the United States, it was George Perkins Marsh in the mid-1800s who first examined human alteration of the environment and then established a modern-day benchmark for inquiry into the topic.[6] A second influential figure, some thirty years Marsh's junior, was John Wesley Powell. A geologist by training, he pioneered interdisciplinary research on the study of land forms and erosion during surveying expeditions in the Great Plains and the Basin and Range regions of the western United States.[7]

Much of the current concern over environmental degradation in tropical areas can be traced back to observations and publications produced as a byproduct of the European colonial period, particularly in the late nineteenth and early twentieth centuries.[8] This body of information was significant in its early recognition of a problem no longer localized but characterized by a larger dimension of national or even global proportions. It also

---

3 For example see Plato's *Critias* with reference to impoverished Athenian soils attributable to deforestation and consequent erosion, which rendered the land unsuitable for growing produce, trees, or retaining water. For a comprehensive account documenting our changing perception of and relationship to nature to the mid-nineteenth century see Clarence J. Glacken, *Traces on the Rhodian Shore* (Berkeley: University of California Press, 1967).

4 Glacken, *Traces on the Rhodian Shore,* p. 342.

5 Fabre noted five downstream issues attributable to the upland destruction of forests and loss of soils. These were the ruination of river bank properties, impacts on downstream navigation, increased strife among individuals formerly separated by in-filled streams, deposition of alluvium and sediment, and diminishment of waters and springs feeding rivers. See Glacken, *Traces on the Rhodian Shore,* pp. 698-702.

6 One of his most notable contributions was to demonstrate that humans, rather than being at the mercy of nature, were in fact increasingly becoming their own agents working independently of nature. See George Perkins Marsh, *From the Earth as Modified by Nature: A New Edition of Man and Nature* (1874; reprint ed., St. Clair Shores, Mich.: Scholarly Press, 1985), pp. 42-43; and William L. Thomas, Jr., et al., eds., *Man's Role in Changing the Face of the Earth* (Chicago: University of Chicago Press, 1955), pp. xxviii-xxix.

7 See John Wesley Powell, *Report on the Lands of the Arid Region of the United States, with a More Detailed Account of the Lands of Utah,* report submitted to the 45th Congress, 2d Session, H.R. Exec. Doc. 73, 1878.

8 For example, soil erosion was viewed as the "gravest danger threatening the security of the white man and the well-being of the colored man in the tropical and sub-tropical lands of Africa and India." See G.V. Jacks and R.O. Whyte, *The Rape of the Earth* (London: Faber and Faber, 1939), p. 20.

specifically identified the human role in contributing to degradation largely through accelerated soil erosion attributed to land clearing, overgrazing, and application of improper agricultural technologies.[9]

Many of these concerns were brought together and viewed from a geographical perspective in a major international symposium in 1955. The symposium was significant in that it marked an early attempt by scientists representing a number of fields in both the social and natural sciences to discuss the nature and significance of our impress upon the earth within an integrated, interdisciplinary framework.[10] This was followed by a second international symposium in 1967, held in response to the growing concern over population pressure and its effect on the use of earth's resources.[11]

During this period of heightened interest in the environment, social scientists focused increasingly on the roles of economics and social equity as factors contributing to the issue. This served to inject into scientific inquiry an ideological perspective divided along Marxist and neoclassicist lines and fueled a debate among academicians and practitioners alike over the fundamental dynamics of global environmental degradation.

Initially, neither Marx nor Engels appeared to have associated environmental decline with the application of capital. Capital was accumulated and wealth created as a result of the exploitation of labor power, and the environment (and natural resources) served only in an enabling role.[12] Similarly, Marxism did not differentiate between the role of capital in developed (metropolitan) and developing (periphery) countries. However, as Marxist development theory evolved, it was argued that for a capitalist economy to be established in the periphery it was necessary to create market conditions based on the country's underlying natural resources and in so doing cause the destruction of the local economy (that is, substitute commodity production for local self-sufficiency).[13] Much of the Marxist argument, buttressed with numerous empirical examples, linked this loss (and its inherent coping strategies) to widespread land degradation.[14] Following

---

9 See Sir Harold Glover, *Soil Erosion* (London: Oxford University Press, 1944), pp. 4-9; Fairfield Osborn, *Our Plundered Planet* (London: Faber and Faber, 1948), pp. 56-58; and Edward Hyams, *Soils and Civilization* (New York: Harper and Row, 1958), pp. 20-26.

10 See Thomas, Jr., et al., eds., *Man's Role in Changing the Face of the Earth*, pp. xxxvi-xxxvii.

11 See Wilbur Zelinsky, Leszek A. Kosinski, and R. Mansell Prothero, eds., *Geography and a Crowding World* (New York, Oxford University Press, 1970), pp. 1-601.

12 Michael Redclift, *Development and the Environmental Crisis* (London: Methuen, 1983), p. 7.

13 Ibid., p. 14.

14 Some of the earliest and best documented accounts linking the excesses of capitalism to land degradation in the developing countries comes from the Sahel, describing how commer-

the revisionism in Marxist development theory during the 1960s and 1970s, proponents began to argue that the global environmental situation was a product of the combination of capitalist underdevelopment in the LDCs and overdevelopment in the world's industrialized countries, the latter characterized by industries' incessant need to promote demand for nonessential consumer goods in an increasingly competitive and overcrowded industrialized world, in order to survive. In addition, poor people were said to impose excessive strains on the carrying capacity of their immediate environment owing to their personal needs (e.g., the need for cash income to repay debts and meet necessities). These needs are imposed on them by structural factors driven by international trade relationships in the global capitalist marketplace. Thus environmental degradation in the developing countries is a "capitalist" problem and the solution is a change in the global economic structure defined by a shift toward increased entitlements for the LDC poor.[15]

In rebuttal, the neoclassicists typically referred to the absence of theoretical underpinnings for natural resources and environmental issues in early Marxist theory. They argued further that capitalist economies do not monopolize the degradation of the environment, typically citing examples from Russia and China. Finally, proponents argued that the market economy, rather than contributing to environmental degradation, in fact provides the means to achieve increased efficiencies of resource use by ameliorating waste in the production process (for instance, a firm's continual search for new and more efficient production technologies to remain competitive).[16]

In the midst of this sometimes contentious debate it is all the more remarkable that the 113 nations attending the 1972 United Nations Conference on the Human Environment reached consensus on a response to the growing problem of environmental degradation and its consequences for the human condition.[17]

---

cialization of the agricultural sector under the colonial powers undermined traditional strategies for coping with drought, putting large sections of the population at risk. Another commonly used example links directly the actions of multinational companies to the despoliation of a host country's environment (e.g., the logging of the tropical forests). See Piers Blaikie, "Natural Resource Use in Developing Countries," in *A World in Crisis*, ed. R.J. Johnston and P.J. Taylor (Oxford: Basil Blackwell, 1986), p. 114.

15 Redclift, *Environmental Crisis*, p. 130.

16 Ibid., p. 113.

17 One of the potentially contentious issues at Stockholm involved the perceived differences between the developed and developing countries concerning the relationship between environment and development. Simply put, many representatives from the developing nations considered global environmental degradation to be an issue of low national priority and a luxury only the developed countries could afford to address. However, partly as a result of a series of meetings held prior to Stockholm that focused on the relevance of environmental considera-

Not so optimistic, however, were the reports originating from the 1972 meetings of the Club of Rome.[18] The club's publication, *Limits to Growth*, served to broaden and diversify both the basis of the aforementioned debate and the protagonists' membership. At the heart of the controversy was the Club of Rome's predictive computer model simulations indicating an increasingly impoverished world as the result of the synergistic effects of high population growth rates, ever-expanding market economies, and finite natural resources. It was concluded that continued growth and its effect on the rate of use of nonreplaceable resources could be sustained and would result in eventual global economic collapse. The report identified three alternative courses of action in response to the problem, recommending self-imposed constraints on growth as the only logical solution.

The model's opponents, sometimes referred to as the "optimists," generally believed that the most important resources were capital, technology, and educated people, which in combination could resolve issues over population growth in a world of finite resources through the use of substitution, new discoveries, and recycling.

By the end of the decade the model's critics appeared to have succeeded in discrediting the Club of Rome's report.[19] Chief among the criticisms responsible for the model's demise were its failure to account for human adjustments and technological advances affecting resource use and depletion, and its incomplete information particularly with regard to the status of resources.

In 1980, the debate was reinvigorated with the publication of the report of the Brandt Commission. Economic stagnation in many of the developed countries together with growing economic disparities between developed and developing countries provided the justification for the creation of the commission. Their report, issued some eight years after Stockholm, was significant in that it explicitly linked the destruction of the natural environment with rural poverty.[20] It was argued that economic growth in the LDCs stimulated by increased international investment would improve the developing countries' capacity to meet their basic hu-

---

tions to the achievement of sustained economic development, many of these concerns were allayed, thus providing the basis for a consensus at the conference. See Robin Clarke and Lloyd Timberlake, *Stockholm Plus Ten* (London: Earthscan, 1982) and United Nations, *Report of the United Nations Conference on the Human Environment* (New York: United Nations, 1973), pp. 1-4.

18 Blaikie, "Natural Resource Use," p. 108.

19 Ibid., p. 108.

20 Independent Commission on International Development Issues, *North-South: A Programme for Survival* (Cambridge: Massachusetts Institute of Technology Press, 1980), p. 79.

man needs and in so doing help reduce poverty and the pressure on the environment.[21]

The Marxists' response to the commission's recommendations advocated deindustrialization and reduction of competition among the developed countries, which in turn would reduce demand for raw materials from the developing countries and thereby lower pressures on natural resources.[22]

Into this debate the international environmental community launched a new initiative with the publication of the World Conservation Strategy (WCS) in 1980. Published in part as a result of Stockholm's failure to achieve many of its goals, the WCS outlined a strategy and means to achieve many of the 1972 convention recommendations. The strategy, a seemingly ideologically neutral document, identified three specific objectives for planning for living resources conservation. These were: (1) to maintain essential ecological processes and life-support systems; (2) to preserve genetic diversity; and (3) to ensure the sustainable utilization of species and ecosystems.[23]

However, despite its well-founded intentions, it was subsequently criticized for failing to acknowledge the existence, much less the findings, of the Brandt Commission.[24] More specifically, it was criticized for ignoring both the social and political considerations associated with the environmental development process.[25]

The lack of consensus over the structural causes of environmental degradation, the Brandt Commission's explicit linking of poverty with environmental degradation, and the criticism of the failure of the WCS (and other international conservation efforts) to address social and political issues amidst increasing indications that many environmental issues in the LDCs have underlying social causes clearly demonstrate the need for new approaches to the study of environmental degradation. In response to this need, a number of recent contributions have proven useful in integrating analytical tools from both the natural and social sciences.

---

21 Ibid., pp. 267-281.

22 Redclift, *Environmental Crisis*, pp. 55-56.

23 International Union for Conservation of Nature and Natural Resources, *World Conservation Strategy* (Gland: International Union for Conservation of Nature and Natural Resources, 1980).

24 Redclift, *Environmental Crisis*, p. 15.

25 This appears to be a failing shared with many of the recent international conservation field research publications as well. These have been criticized for their overemphasis on natural phenomena and our impact on same at the expense of "human situations" leading to environmental degradation. See Harold Brookfield, "On Man and Ecosystems," *International Social Science Journal* 34 (September 1982): 375-376.

Integrative Approaches

It has been argued that geography's unique contribution to science has stemmed from its role in integrating its four major research themes, spatial, environmental, regional, and physical analysis.[26] In view of this interpretation of geography's mission and the aforementioned needs and justification for interdisciplinary inquiry, a brief review of the field's man/land tradition and its more relevant contributions is warranted.

During the period between the writings of Marsh and Powell geography was being formally institutionalized as an academic discipline. Departments were first established in several European countries, and early research was dominated by physiographic studies reflecting the natural sciences background of many of the field's founders.[27] However, this initial emphasis began to shift to accommodate a human element partly as a means of avoiding the dominance of the field by one academic discipline (i.e., geology). As a result geography expanded, and the field's man/land tradition was initiated.[28] While this shift first began in Europe by the turn of the century, it was the geographers in Britain and the United States who came to dominate the newly expanded field.[29]

This early focus on humans and the environment eventually led to a line of inquiry supporting the concept that environment was the primary determinant of human behavior (environmental determinism).[30] The extremism of this view led to a backlash both within the field and among other disciplines, resulting in the formulation of a number of rival doctrines, such as possibilism, and new lines of inquiry all together, such as

---

26 While perhaps not universally accepted, this view has been supported in and by a number of prominent forums and individuals. See Ad Hoc Committee on Geography, *The Science of Geography* (Washington, D.C.: National Academy of Sciences–National Research Council, 1965), pp. 2-10; Peter Haggett, *Geography: A Modern Synthesis* (New York, Harper and Row, 1972), p. xiii; and Brian J.L. Berry, "Approaches to Regional Analysis: A Synthesis," *Annals of the Association of American Geographers* 54 (March 1964): 2-3.

27 Similarly geography in the United States owed its beginnings largely to the work of natural scientists, particularly geologists. For example see N.S. Shaler, "The Economic Aspects of Soil Erosion," *National Geographic Magazine* 7 (1896): 328-338.

28 Carl O. Sauer, "Foreword to Historical Geography," *Annals of the Association of American Geographers* 31 (March 1941): 2.

29 As an example, in the country's first geography department at the University of Chicago it was the stated function of the department to provide courses that could serve as a bridge between the natural and social sciences. Cited in Marvin Mikesell, "Geography as the Study of Environment: An Assessment of Some Old and New Commitments," in *Perspectives on Environment*, ed. Ian R. Manners and Marvin W. Mikesell (Washington, D.C.: Association of American Geographers, 1974), p. 2.

30 For an example illustrating this perspective see Ellsworth Huntington, *Civilization and Climate*, 3d ed. (New Haven: Yale University Press, 1915).

spatial analysis.[31] Despite these inauspicious beginnings there have been a number of seminal contributions over the intervening seventy years that are relevant to the research topic. In the early 1920s Barrows defined geography as a kind of human ecology centered on the study of the relationships between human activities and the environment, and he argued the need to focus inquiry on the adjustment of people to the environment.[32] His writings provided much of the basis for the next significant stage in the tradition, the contribution of Carl Sauer and his students, who advocated the significance of human-induced modifications to the landscape. This in turn led a number of researchers to study issues such as deforestation, soil erosion, and reclamation.[33]

One line of inquiry that evolved from these early twentieth-century foundations was the study of environmental perception, a field of inquiry that has been cited as one of the few examples of successful integration of social and ecosystem modeling.[34] Many of these perceptual studies stemmed from G.F. White's early research on natural hazards. Beginning with a focus on the impact of government sponsored flood-control works on property and life in the United States, the field expanded and diversified to include the study of human perception and adjustment to hazards, both natural and artificial, and theories to explain observed responsive behavior.[35]

One particularly relevant contribution from White and his colleagues was an international series of comparative case studies designed to test a number of hypotheses using a diverse range of hazards from a variety of settings. The effort was successful in expanding and diversifying the field of inquiry, but the perception and behavior components were sharply criti-

---

31 However it has been noted that there existed ample qualification in the determinists' publications, many of whom were merely attempting to search for standards of correlation and measurement between climatic change and human consequences of different habitats. See Mikesell, "Study of Environment," pp. 2-3.

32 Barrows posed two problems in need of study within the discipline of regional economic geography: how and why do we use our land, and what are the advantages and disadvantages of the region for our use? He noted that the study must include an examination of the features of the cultural landscape (i.e., surface manifestations of human occupation of the region) and how these have changed over time. Harlan H. Barrows, "Geography as Human Ecology," *Annals of the Association of American Geographers* 13 (March 1923): 1-14.

33 See Carl O. Sauer, "The Morphology of Landscape," *University of California Publications in Geography* 2 (1929): 37-53; and Mikesell, "Study of Environment," p. 4.

34 Anne Whyte, "The Integration of Natural and Social Sciences in the MAB Programme," *International Social Science Journal* 34 (1982): 423.

35 Three seminal contributions to geographic science attributed to White were: the concept of human adjustment, its broad and theoretical range, and the role of perception and decision-making in resource management. See Robert W. Kates and Ian Burton, eds., *Geography,*

cized. Criticisms included: failure to account for differences in cultural and societal organization in the use of a standardized interview format in a multicultural survey; exclusion of key participants in the survey; and failure to address the political and economic processes and forces that might accentuate the gravity of hazards.[36]

In the late 1950s and early 1960s there was a renewed interest among many researchers to find a more effective means to analyze ecological and social systems within a common framework. Both determinism and possibilism were judged inadequate for precise analysis, the former for exaggerating the causative nature of the environment, the latter for the relegation of the environment to a passive role in the development of human culture. Initial efforts to apply ecological principles to human populations treated society like any other biotic phenomenon. However in its simplest form, this approach to human ecology came to be regarded as a reductionist use of ecological principles, which resulted in investigations more akin to locational theory than inquiry into the relationships between the environment, social, and cultural processes.[37]

One response to these criticisms was to limit the application of ecological principles to selected aspects of human social and cultural life. An example was Geertz's 1963 comparative study of the effects of agricultural intensification under different management systems in Indonesia. Building on Steward's earlier principles of cultural ecology, Geertz's approach required the discrimination of two sets of variables related through the processes of energy interchange between people and their surroundings; these were a population's cultural core and the related environmental variables.[38] These two sets were considered reciprocal in nature and once one was discriminated, it became possible to determine the second through correlative analysis.[39] Intervariable integration, it was argued, came from subsequent analyses and viewing the system from a functional perspective.

Resources, and Environment (Chicago: University of Chicago Press, 1986), vol. 1, *Selected Writings of Gilbert F. White*, p. xii.

36 See William I. Torry, "Hazards, Hazes and Holes: A Critique of the Environment as Hazard and General Reflections on Disaster Research," *Canadian Geographer* 23 (Winter 1979): 371-372; and Eric Waddell, "The Hazards of Scientism: A Review Article," *Human Ecology* 5 (March 1977): 75.

37 See Clifford Geertz, *Agricultural Involution* (Berkeley: University of California of Press, 1963), pp. 5-6.

38 The cultural core is defined as "the constellation of features which are most closely related to subsistence activities and economic arrangements . . . and includes . . . social, political and religious patterns as are empirically determined to be closely connected with these arrangements." See Julian Steward, *The Theory of Culture Change* (Urbana: University of Illinois Press, 1955), p. 37.

39 Geertz, *Agricultural Involution*, p. 8.

In a later, much cited book, Rappaport described his research on the significance of ritual in human ecology for a New Guinean tribal group.[40] The study was important in its advocacy of an ecological approach in which populations and ecosystems should be considered the basic units of analysis whose discrimination, owing to the continuity of natural phenomena, must be based on criteria established on strategic grounds in support of the research.[41] Rappaport rejected the use of culture as an analog for animal populations. He argued that human ecology should be "synthetic" in nature, requiring as a first step the definition of terms common to all populations, prior to defining distinguishing properties (such as culture) among populations.[42]

In a different approach to the problem, UNESCO developed the concept of the human-use system, first proposed by a task force studying the contribution of social sciences to the organization's Man and the Biosphere Programme (MAB).[43] The concept was based on the identification of a "parallel" social-based system through which natural ecosystems are managed. It was argued that the recognition of these two systems, the natural and the social frameworks in which natural ecosystems are managed, and the study of their relevant intersections would facilitate interdisciplinary inquiry.

The UNESCO approach has since been criticized for its inadequacy in facilitating rigorous problem selection and for integrating macrolevel and microlevel studies as well as social with natural science inquiries.[44] As an alternative, the concept of a human-use ecosystem was proposed. This concept lowered the scale of analysis to specific ecosystem-population interactions (for instance, the interaction between a coral reef and fishing community), which would more readily facilitate the integration of natural and social sciences.[45]

Suffice it to say, each framework has made its share of contributions to an increased understanding of the complexity of the human environment and facilitated further interdisciplinary inquiry. It is also apparent that there is no single integrative framework universally recognized as fulfilling

---

40 Roy A. Rappaport, *Pigs for the Ancestors*, 2d ed. (New Haven: Yale University Press, 1984).

41 Ibid., pp. 380-388.

42 Ibid., pp. 383-384.

43 See United Nations Educational, Scientific and Cultural Organization, *Task Force on the Contribution of the Social Sciences to the MAB Programme: Final Report* (Paris: UNESCO, 1974).

44 See Harold Brookfield, "On Man and Ecosystems," *International Social Science Journal* 34 (September 1982): 376.

45 Ibid., p. 389.

the demands of such inquiries into the problem. This would seem to lead to the conclusion that the development of these integrative frameworks is, and will likely be, a continuing, dynamic process partly shaped by the needs of the particular research problem.

The Historical and Geographical Context

The arguments for framing the study of environmental degradation within its historical context are relatively obvious. At one level, there must be a recognition of the lag that exists between cause and effect. This can occur at several levels (e.g., the temporal lags between devegetation and soil erosion or policy implementation and changes in land use). At another scale, there is the realization that landscapes are continually being modified by biophysical processes over time. This in turn, frames one of the questions central to the inquiry, namely, what has been the human role in affecting these processes over time?[46]

Not surprisingly, these considerations have long been recognized by geographers. For example, Barrows stated that human adjustments to the environment in the present represent only a stage in a process, and he suggested that the study of the historical past was critical to understanding the present.[47] In another example, G.F. White cited the importance of approaching water management from a historical perspective, noting that management of an area evolves through historical stages characterized by sets of distinct problems and their respective resolutions and that, for purposes of anticipating future planning constraints and needs, an understanding of the past is essential.[48]

With respect to the geographical context, White in an earlier paper identified a number of critical elements recommended for consideration in decision-making in water management.[49] One of these, social guides, he defined as "the character of constraint or encouragement which society gives through its institutions of custom, attitude, education, and organization [to the other five aspects of decision-making]."[50]

---

46 Blaikie and Brookfield, introduction to *Land Degradation,* pp. xx-xxi.

47 See Barrows, "Human Ecology," p. 12.

48 See Gilbert F. White, "Role of Geography in Water Resources Management," in *Man and Water*, ed. Douglas James (Lexington: University of Kentucky Press, 1974), pp. 102-105.

49 These were range of choice, resources estimates, technology of water management, economic efficiency, spatial linkages, and social guides. See Gilbert F. White, "Contributions of Geographical Analysis to River Basin Development," in *Readings in Resource Management and Conservation*, ed. Ian Burton and Robert W. Kates (Chicago: University of Chicago Press, 1963), p. 381.

50 Ibid.

The aforementioned are just a few examples illustrating the importance geographers have placed on incorporating this perspective into the man/land tradition.

Applications to the Coastal Watershed

Interdisciplinary field studies of upstream/downstream aspects of environmental degradation are few. Until recently sociocultural issues, particularly with regard to impacts on local populations, have been one of the least understood and most lightly regarded aspects of water resources development.[51] There was considerable interest in the study of the river basins as unified physical entities in the 1960s and 1970s, and a number of interdisciplinary research activities were sponsored through coordinated international scientific programs.[52] However, many of these studies were oriented primarily toward research on biophysical processes.

As awareness grew of the importance of the social dynamics of river basin development, researchers became increasingly interested in the subject. By 1984 researchers were proposing that watersheds were the most appropriate spatial unit for many kinds of regional rural development planning activities in developing countries, capable of bringing together the myriad aspects of both natural and social processes.[53] In at least one paper during this period, the river basin's social interdependencies along an upstream/downstream gradient were implicitly, if not explicitly, identified.[54] Of even greater relevance to the present paper's theme was the recognition that social issues in the river basin were not confined solely to the domain of large-scale, capital-intensive development projects, but could also be attributed to the rapid expansion of small-scale agriculturalists in upland areas. Specific impacts identified included modifications of the hydrological cycle, increased rates of soil erosion, reduction in genetic diversity, and losses in supply of available protein related to reduced levels of indigenous

---

51 See Ludwik A. Teclaff, "Harmonizing Water Use and Development with Environmental Protection," in *Water in a Developing World*, ed. Albert A. Utton and Ludwik Teclaff (Boulder: Westview Press, 1978), p. 81; and James L. Wescoat, Jr., "The 'Practical Range of Choice' in Water Resources Geography," *Progress in Human Geography* 11 (March 1987): 43.

52 These included the International Geophysical Year, UNESCO's MAB Program, the International Hydrological Decade, and the establishment of the Scientific Committee on Problems of the Environment (SCOPE).

53 See Lawrence S. Hamilton and Peter N. King, "Watersheds and Rural Development Planning," *Environmentalist* 7 (1984): 80.

54 It was noted that successful development activities within a river basin planning framework required the support, or at least acquiescence, of persons and groups outside the immediate planning area subject to downstream project-related effects (low flows, reduced water quality, etc.). See ibid., pp. 80-86.

wildlife populations.[55] Perhaps of equal relevance was the belated recognition of the interplay between the biophysical components and human managers of these upland agroecosystems in contributing to downstream environmental disturbance.[56]

Despite these advances little appears to have been published about effects on coastal communities. One exception is a 1985 study that attempted to document the cause-and-effect relationships of accelerated soil erosion in Kenya's Athi River basin to coastal beaches and coral reef degradation. While the study did not treat the social dynamics leading to conflict, it nevertheless contributed a better understanding of the problem, and laid much of the foundation for the present study.[57]

A second exception is a remarkable, twenty-nine-part technical report series addressing the marine and terrestrial ecosystems of the U.S. Virgin Islands National Park and Biosphere Reserve. Specific reports based on field studies addressed historical land use, soil and vegetative characteristics, and sedimentation and reef development within common watersheds in the study area.[58]

The Case Study

An appropriate-sized watershed would appear to be an ideal area to test one or more existing integrative frameworks. First, the boundaries are clear. The concept of watershed as a natural ecosystem with a well-defined physical boundary has been widely recognized and adopted in a number of ecosystem studies.[59] The definition of the second unit of analysis, the human population, would consist of those individuals who occupy a common watershed and are affected chiefly by a resource use conflict attributed to modification of the physical environment.

The proposed definition of these units appears to be most compatible with Rappaport's concept of an ecological population. Moreover, he recognizes the distinction between local and regional populations, defining the

---

55 See Terry Rambo, "Human Ecology Research on Tropical Agroecosystems in Southeast Asia," *Singapore Journal of Tropical Geography* 3 (June 1982): 88.

56 Ibid., p. 89.

57 Random DuBois, "Catchment Land Use and Its Implications for Coastal Resources Conservation in East Africa and the Indian Ocean," *Ocean Yearbook 5*, ed. Elisabeth Mann Borgese and Norton Ginsburg (Chicago: University of Chicago Press, 1985), pp. 221.

58 For a synthesis of these studies see Caroline S. Rogers and Robert Teytaud, *Marine and Terrestrial Ecosystems of the Virgin Islands National Park and Biosphere Reserve* (St. Thomas: U.S. National Park Service, 1988), p. 109.

59 See F. H. Bormann and G. E. Likens, "The Watershed-Ecosystem Concept and Studies of Nutrient Cycles," in *The Ecosystem Concept in Natural Resource Management*, ed. George M. Van Dyne (New York: Academic Press, 1969), pp. 49-78.

latter as "more dispersed systems of intraspecies exchanges with the most important transactions likely to be exchanges of personnel, information, and services of various sorts, as well as trade goods and valuables."[60] In cases where spatial limitation is poorly developed (such as weak territoriality) he has suggested that the ecological population may have to include all groups living in the region, and that the region as a whole be considered the ecosystem (for purposes of the study two groups are identified, upland and lowland residents inhabiting the common watershed). Finally he recognizes that even where regional systems and ecosystems are geographically and demographically coextensive, the analysis of how these systems are articulated can be productive.[61]

Framing the study within the proper historical and geographical context is particularly important in the case study. Many present-day land-use patterns and characteristics in the Philippines can be traced directly back to early Spanish traditions. Moreover, Siquijor's physical and climatic constraints (and the Central Visayas generally) appear to exert a crucial influence on the development of livelihood systems.

---

60 Rappaport, *Pigs for the Ancestors*, p. 390.

61 Interestingly, he notes that one question worthy of study is the extent to which one system affects events in the other (e.g., ecosystemic processes are buffered against economic processes) or whether one system is continuously, sporadically, or periodically responsible for occurrences in the other. Ibid., p. 392.

Chapter 3

THE GEOGRAPHICAL CONTEXT

Introduction

There exist few countries better suited than the Philippines for studying the effects of nonsustainable land-use practices in upland ecosystems. In part, this can be attributed to the colonial policies that favored the development of a large plantation-based agrarian economy.[1] These enterprises characteristically were located in the highly productive lowlands where fertile soils predominated. This historical legacy contributed to the evolution of a modern-day oligarchy whose sources of wealth can be traced to ownership of large agricultural estates, many of which are still intact. The existence of large landholdings in the hands of a few, together with the country's high population growth rate, continues to serve as a major driving force behind upland occupancy and the rapid exploitation of these natural resources.[2]

Physiography

In economic terms the islands of the Central Visayas, after northern Mindanao, are perhaps the poorest areas in the Philippines. Although the explanation for the existing economic conditions is a complex one, topo-

---

1 D.J.M. Tate, *The Making of Modern South-East Asia*, vol. 2, *Economic and Social Change* (Kuala Lumpur: Oxford University Press, 1979), pp. 425-435.

2 Historically, upland settlement was encouraged by government resettlement through homestead programs extending as far back as 1781. However most evidence indicates that the present situation of intense and widespread exploitation of upland marginal lands, characterized by extreme slopes, poor soils, and highly erosive rains, is largely a post–World War II phenomenon. Unplanned occupancy of these areas has resulted in a number of well-documented effects in the Philippines, including the occupation of public lands, loss of natural forests, and widespread soil erosion. Nevertheless there appears to be little information regarding the downstream effects specifically attributable to the farming of marginal lands. See C.J. Cruz, "Demographic Issues in Upland Development," paper presented at the Workshop on a National Strategy for Sustainable Development of Forestry, Fisheries, and Agriculture, Manila, Philippines, March 30-31, 1987, p. 6.

graphical constraints, poor soils, and high population growth rates typical of the region's principal islands are major underlying factors.[3] In many ways Siquijor, a small island with an estimated area of 290 km^2, typifies the issues, constraints, and driving forces that define the Central Visayans' relationship with their environment.[4] The island, located near the southern tips of Cebu and Negros Occidental (9° 10' N, 123° 35' E), is bounded by Negros Oriental to the west, Cebu and Bohol to the north, the Bohol Sea to the east, and Mindanao to the south (figure 1).

Siquijor's terrain can best be described as rugged, with some 36 percent of the island characterized by slopes exceeding 25°.[5] Drainage is radial, originating in the central highlands and formed of numerous steep and short river basins. Mt. Malabahoc at 605 m is the island's highest peak and dominates the steep-sloped, highly dissected central hills characteristic of the upland areas. Coastal lowlands are dominated by a rolling landscape typified by flat, well-dissected terraces, hills, and in-filled valleys that terminate abruptly in narrow coastal alluvial plains typically set off by sea cliffs.[6]

The island appears to be geologically detached from the other Central Visayan islands owing to its well-defined separation by deep submarine slopes. Siquijor's geological history can be briefly summarized in five discrete steps. These are: (1) an accumulation of volcanics attributable to submarine eruptions; (2) a slow uplift of the ensuing volcanic debris associated with reef formation and the deposition of shale surrounding the reef complex during the upper Miocene; (3) the slow emergence of this formation above sea level often causing reefal material to off-lap and cover the adjacent shale formation; (4) rapid subsidence of the insular mass resulting in a second reef formation; and (5) a second, more gradual reemergence of the mass contributing to the island's characteristic terrace formations.[7]

As a result of the geological evolution of the island Siquijor's lithology is characterized by marine sediment deposits in the form of shale, both

---

3 A study conducted by the World Bank identified the Central Visayas region (Region 7) as the second poorest in the country based on the application of a poverty index measured along guidelines established in an earlier 1975 survey. See World Bank, *Aspects of Poverty in the Philippines: A Review and Assessment*, vol. 2 (Washington, D.C.: World Bank, 1980), pp. 8-10.

4 Republic of the Philippines, National Economic and Development Authority, *Central Visayas Five-Year Development Plan, 1978-1982* (Cebu City: National Economic Development Authority, 1977), p. 1.

5 Republic of the Philippines, Siquijor Provincial Development Staff, *Capital Development Plan* (Siquijor: Office of the Governor, n.d.), no pagination.

6 Republic of the Philippines, Bureau of Mines, *Preliminary Report of the Geology and Manganese Deposits of Siquijor Island, Negros Oriental, Philippines*, by Ronald K. Sorem (Manila: Bureau of Mines, 1951), p. 3.

7 Ibid., pp. 17-18.

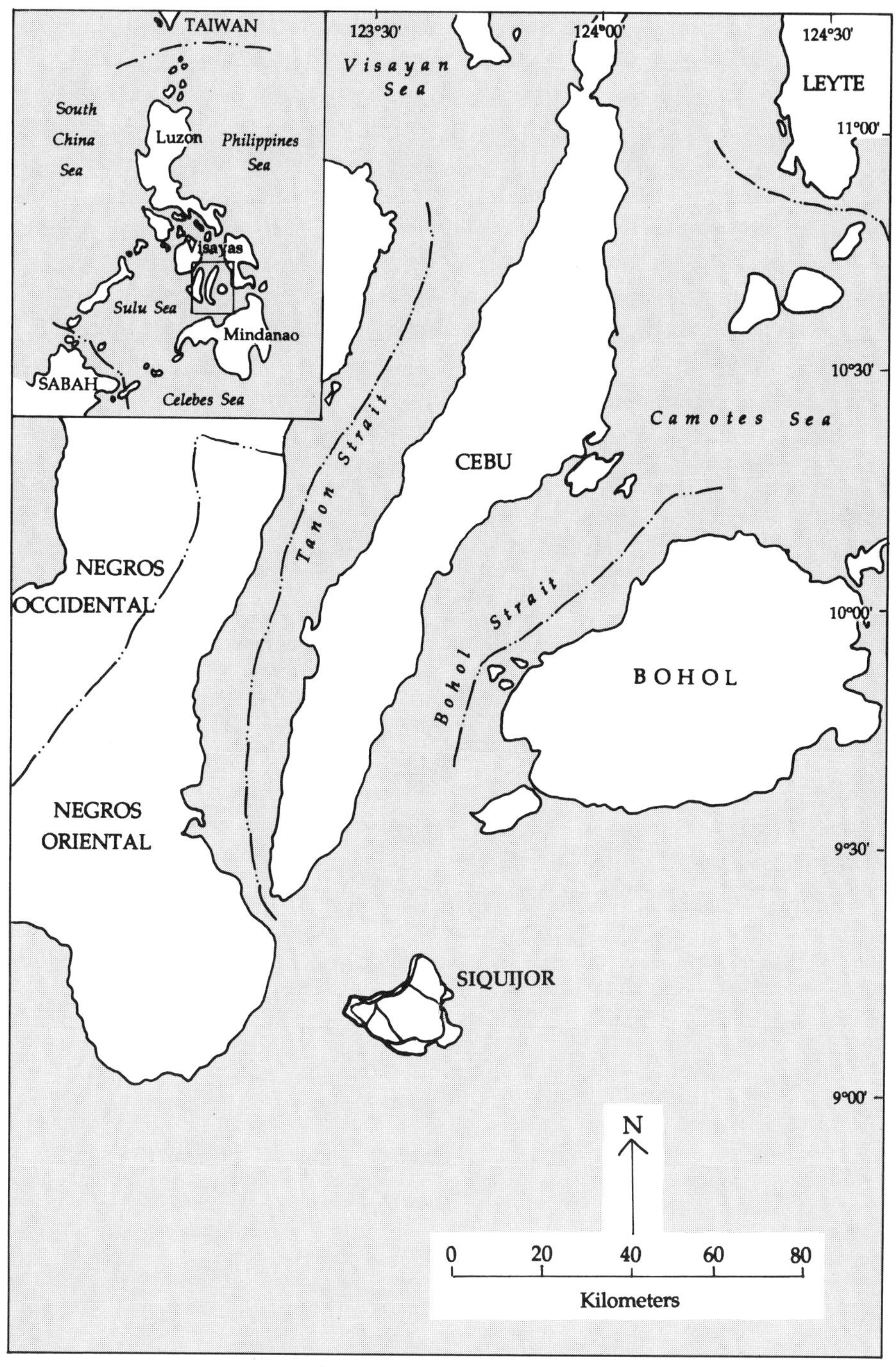

Fig. 1. Central Visayas region, Republic of the Philippines

calcareous and noncalcareous, and limestone. The weathering of these materials has resulted in soils dominated by clays comprising one of four series, Lugo, Bolinao, Faraon, and Mandawe.[8] In addition to these soils, volcanic and hydrosoils are known to occur to a significantly lesser extent.[9]

The island has lost most of its natural forests. The only significant remaining cover is represented by forests planted in the 1930s, providing a semblance of watershed protection and source of building materials but in aggregate covering only 3 percent of the total insular area.[10]

Although there is no definitive evidence of the instrument of devegetation, it has been suggested that *kaingin* farming (shifting agriculture) has been the primary cause.[11] Even less clear is when the island lost most of its primary forest.[12] Regardless of the specific point in time, by 1938 Siquijor's

---

8 The calcareous shale-derived soils belong mainly to the Lugo soil series. This is a relatively productive upland soil typical of hilly topography and narrow valleys; its characteristic differential weathering contributes to sharply contrasting relief typical of much of the Central Visayas. Upland limestone-based soils termed Bolinao (Terra Rossa) and Faraon (Rendzina) in the Philippines are also considered moderately fertile though they are rarely very deep (10-20 cm) because of the parent material's solubility and rapid weathering. Unlike the shale-based soils, the latter are typically stony and often occur as outcrops where the thin surface layers have been eroded. Because of their porous nature these soils are poorly suited for lowland rice; as a result corn, coconuts, bananas, and citrus fruits are the crops of choice. Mandawe, the fourth soil type, is a lowland soil found in older lowland plains and/or terraces with poor internal drainage and low permeability. See Frederick L. Wernstedt and J.E. Spencer, *The Philippine Island World* (Berkeley: University of California Press, 1967), p. 69; and Republic of the Philippines, Department of Agriculture and Natural Resources, Bureau of Soils, *Soil Survey of Negros Oriental*, by Alfredo Barrera and Jose Jaug, Soil Report no. 26 (Manila: Bureau of Printing, 1960), pp. 76-77.

9 Republic of the Philippines, Siquijor, *Capital Development Plan,* unpaged.

10 Evidence of original forest is fragmentary. One of Siquijor's earliest recorded names was Katugasan, which is the common name for a species of hardwood commonly known as molave (a diptocarp species), indicating that this may have been a predominant species during the Spanish period. Additional evidence may be found in small extant forest patches, observations of materials used for construction of houses standing from the last century, and oral histories indicating their occurrence locally.

11 Republic of the Philippines, Department of Agriculture and Natural Resources, *Soil Survey*, p. 14.

12 In 1907 Siquijor was described as "most favored with forest resources." However accounts from the late 1800s derived from early museum-sponsored expeditions described Siquijor's upland areas as "little better than a limestone rock" (1899) and its hills as "mere masses of coral rag, to which a few trees cling" (1898). While earlier accounts failed to mention Siquijor specifically, descriptions from the neighboring islands varied with regard to the quality of their forests, from considering them an important resource ("mayor riqueza forestal y major variedad") in the islands of Leyte and Negros, to lamenting their degradation ("lamentable decadencia") in Cebu. One report of Cebu in 1902 claimed that "almost every stick of merchantable timber has been cut away." In contrast, deforestation in Negros appears to have lagged somewhat, as some 50 percent of the original rain forest was estimated to have remained until the early 1940s. At present there is little more than 5 percent of primary tropical rain forest remaining on the island. See U.S., War Department, Bureau of Insular Affairs,

predominant vegetation appeared to be cogon grass (*Imperata cylindrica*), as evidenced by its presence in the island's principal protected watershed of Mt. Bandilaan and which, with scrub and small remnants of secondary forest, continues to dominate the vegetative cover today.[13]

Siquijor's climatic regime is influenced by its interior position within the archipelago, sheltered in part from moisture-bearing air masses. The island's precipitation pattern (and most of the Central Visayas region) is typified by reduced rainfall and commonly drought from February through April. Rainfall maximums occur with the advent of the southwest monsoon, approximately from June to October.[14] Based on Siquijor's few available records from a station located at one of the island's highest elevations (450 m), estimated annual precipitation is 1,811 mm. Based on these characteristics Siquijor can be considered to be representative of Thornthwaite's sub-humid class C. Koppen's classification system appears less applicable to the Philippines because of its limitations in distinguishing rain forest and monsoon classes in detail, and its failure to consider drought duration, limiting itself to the single driest month.[15]

Report of the Bureau of Forestry of the Philippines Islands, Appendix J., Report of the Chief of the Forestry Bureau for the Period from July 1, 1901 to September 1, 1902 (Washington, D.C.: Government Printing Office, 1902), p. 470; U.S., War Department, *Eighth Annual Report of the Philippine Commission, Part 1* (Washington, D.C.: Government Printing Office, 1907), p. 420; Dean C. Worcester, *The Philippine Islands and Their People* (London: Macmillan Co., 1899), p. 289.; Idem, "Contributions to Philippine Ornithology," *Proceedings of the U.S. National Museum* 20 (1898): 581; Nueva Espana, *Guia Oficial de Filipinas* (Manila: Ramirez y Giraudier, 1884), p. 683; R.C. McGregor, *A Manual of Philippine Birds* (Manila: Bureau of Science, 1909), part 2; F.S. Bourns and D.C. Worcester, "Preliminary Notes on the Birds and Mammals Collected by the Menage Scientific Expedition to the Philippine Islands," *Occasional Papers of the Minnesota Academy of Natural Sciences* 1 (1894): 1-65; and Walter C. Brown and Angel C. Alcala, "Comparison of the Herpetofaunal Species Richness on Negros and Cebu Islands, Philippines," *Silliman Journal* 33 (January 1986): 75.

13 Cogon is a native grass that occurs throughout Southeast Asia and is widely associated with burning of grasslands; its fire resistance is attributed to underground rhizomes and high regenerative capacity. See Commonwealth of the Philippines, Department of Agriculture and Commerce, Bureau of Forestry, *Annual Report of the Director of Forestry of the Philippines* (Manila: Bureau of Printing, 1940), p. 47; and J.L. Falvey, "*Imperata cylindrica* and Animal Production in South-East Asia: A Review," *Tropical Grasslands* 15 (March 1981): 52-56.

14 Republic of the Philippines, National Economic and Development Authority, *Central Visayas*, p. 1.

15 Thornthwaite's system would also seem to be compatible with a local classification scheme based on empirical studies categorizing the Filipino climate into four types, of which type 3 most nearly describes the climate of Siquijor, that is, an absence of very pronounced seasons, characterized by a relatively dry period from November to April and a wet period during the rest of the year without significant rain maximums, along with areas partly sheltered from northers and trades, and open to the southwest monsoon or at least frequent cyclonic storms. See Wernstedt and Spencer, *Philippine Island*, pp. 58-61; and B.S. Lomotan, "Climatic Types of the Philippines," in *Philippine Recommendations for Corn 1970-1971* (Los Banos: Philippine Council for Agricultural Research and Development, 1970).

Surface runoff is rapid, which has contributed to the highly dissected character of the upland areas.[16] Most streams however are ephemeral, and the island has few perennial rivers, the principal ones being the Maria, Senora, and Tag-ibo.

Groundwater is thought to be significant both in terms of presence and abundance, occurring at relatively shallow depths near the coast.[17]

Erosion is both severe and pervasive, particularly in the steep highland areas of the island.[18] This appears to be due to a number of factors. These include the dominance of clays in the upland soils and their characteristically excessive surface but slow-to-moderate internal drainage, susceptibility of shale-derived soils to mechanical erosion, reduced vegetative cover, sharp relief, high-intensity rainfall, and poor land-use practices.

The island's coast, measuring some 72 km in circumference, is characterized by numerous broad alluvial flood plains fronted by small sand beaches typically separated by rocky headlands or the occasional mangrove-dominated estuary. A shallow submarine shelf supports reef patches and marine grassbeds. An exception to this characterization occurs off the island's northwest coast where the shelf widens and supports an extensive coral reef flat. The reefs appear to be in poor condition, partly attributable to the effects of sedimentation.[19]

Surface currents are driven by the prevailing local winds and flow primarily in a westerly direction.[20] In the nearshore areas these wind-

---

[16] Republic of the Philippines, Department of Agriculture and Natural Resources, *Soil Survey*, p. 55.

[17] See Republic of the Philippines, National Water Resources Council, *Rapid Assessment of Water Supply Sources: Province of Siquijor* (Manila: National Water Resources Council, 1982), pp. 1-11.

[18] Based on a 1983 land classification project approximately one-half of the island's lands were classified as steep lands with excessive to very severe erosion. See Republic of the Philippines, Department of Agriculture and Food, Bureau of Soils and Water Management, Agricultural Lands Management Evaluation Division, *Soil/Land Resources Evaluation for Agriculture: Province of Siquijor, Region 7* (Manila: Bureau of Soils, 1985).

[19] A nationwide survey of Philippine coral reefs using living coral cover as a criterion of status rated approximately 71 percent of Siquijor's coral cover as fair or poor. Categories were excellent (75-100 percent live cover), good (50-74.9 percent), fair (25-49.9 percent), and poor (0-24.9 percent). Compared to the study average derived from the other twenty-six sites, Siquijor was slightly worse off than other locations in the Visayas and better off when compared to the average of sites in Luzon and Mindanao. See E.D. Gomez, A.C. Alcala, and A.C. San Diego, "Status of Philippine Coral Reefs," in *Proceedings of the Fourth International Coral Reef Symposium*, ed. Edgardo Gomez et al. (Quezon City: Marine Sciences Center, University of the Philippines, 1981), pp. 275-282.

[20] Republic of the Philippines, Department of National Defense, Coastal and Geodetic Survey, *Philippine Coastal Pilot, Part 1: Sailing Directions for the Coast of Luzon, Mindoro and Visayan Islands*, 5th ed. (Manila: Coastal and Geodetic Survey, 1968), sect. 1, pp. 33-35.

driven currents are heavily influenced by the tidal regime and local physiographic features.[21]

Insular Economy, Population, and Land-Use Practices

Understanding Siquijor's existing land-use patterns, particularly those contributing to erosional processes, requires a brief review of the island's agriculturally dominated economy over time.

The agriculture of pre-Hispanic Siquijor appears to have been characterized by shifting cultivation. The islands were sparsely inhabited, with much of the population located along the coast. Coralline soils typical of the Central Visayas were not conducive to rice production, and millet seems to have been the preferred crop. The island's first known contact with the Spanish came during the 1565 Legazpi Expedition when Esteban Rodriguez led a small exploratory party from the base camp in Bohol to nearby islands. However it was not until 1783 that Siquijor parish was first established. This was soon followed by Can-oan (Larena), Tigbawan (Lazi), Makalipay (San Juan), and Cangmeniac (Maria). With the exception of Enrique Villanueva, all the island's settlements had been established by 1877.

Corn, in all likelihood established for purposes of enlarging agricultural production to support growing Spanish settlements, was first introduced from Mexico in the late 1560s or mid-1570s. By the 1700s the crop had spread widely throughout the region, in part aided by new technologies in the form of home-milling facilities, and it provided the means of supporting the relatively rapid population growth characteristic of the region after 1800.[22] Corn is the mainstay of the subsistence farmer in Siquijor, which surpasses all other islands of the Central Visayas, with an estimated 86 percent of the population dependent on its production.[23] Upland cropping practices are dominated by monocropping on overly steep slopes, contributing to the erosion problem.

Paddy production is of secondary importance to the subsistence economy and is confined largely to the island's coastal and few inland alluvial valleys.

---

21 The tides of the Philippines are affected more by the declination of the moon (tropic tides) than position of the sun and moon respectively. As a result tidal patterns vary from two discernible tides a day occurring within a few days of the moon crossing the Equator to a semidiurnal tide when the moon reaches its greatest declination (north or south) before returning to the diurnal condition some two to five days later. This in turn affects the strength and pattern of coastal currents. Ibid., pp. 1-32.

22 J.E. Spencer, "The Rise of Maize as a Crop Plant in the Philippines," *Journal of Historical Geography* 1 (January 1975): 1-16.

23 Republic of the Philippines, National Economic and Development Authority, *Central Visayas*, p. 36.

A third important subsistence sector is fishing, where most of the catch is consumed locally, with few exports. The annual catch is estimated to be a little over 1,000 mtons. The fishery itself is dominated by municipal fishermen using unmotorized *bancas* (small wooden boats with outriggers), of which there are an estimated 2,800 on the island.

The island's early commercial history was dominated by agricultural exports, as is the case today. An economy dominated by export crops did not begin to develop until after 1750, since much of the interest of the early Spanish in the Philippines was focused on the trade between China and the Americas. As this trade began to erode with the rise of British and American expansionism, the Spanish increasingly emphasized developing a cash-crop economy overseas. The Spanish introduced indigo, pepper, sugar, and tobacco, of which the latter three became widely grown within a few decades. While the higher soil fertility of the island of Negros allowed it to become an exporter of sugar, tobacco appears to have been the only significant commercially grown crop introduced into Siquijor during this period. Other export crops were the indigenous abacá and coconut. Abacá was produced for export in the late 1700s and early 1800s to meet American demands for Manila hemp.[24] However, by the turn of the twentieth century exports came to be dominated by coconut (copra), chickens, and eggs, contrasting with declines in tobacco and abacá (table 1).[25] By the 1980s however the province had become a net importer of corn, unhusked rice (*palay*), and other grains, and its only significant exports were live animals, copra, and low-priced commodities such as cassava chips and peanuts. To meet local food needs Siquijor imported 1,966 and 2,451 mtons of *palay* and corn respectively to complement 1982 production of some 5,760 mtons of grain. The absence of interisland trade figures in the early censuses precludes determining when importation began. However, in the 1918 census Siquijor corn production was estimated to be 29,831 *cavans*. It has been calculated that 3.65 *cavans* per year of hulled rice, corn, or both are required to support an individual. Based on these figures and Siquijor's 1918 population of 56,774 some 207,225 *cavans* would be required of corn, rice, or some combination of both to support the island's basic food needs. The resulting sizable shortfall, one unlikely met by local paddy production, would seem to indicate that staples were imported at the time.[26]

---

24 Tate, *Modern South-East Asia*, vol. 2, pp. 421-496.

25 In addition to these crops the island has been described as an important source of racing horses for the Central Visayas. See Dominador Z. Rosell, "Siquijor Island," *Philippine Magazine* 35 (June 1938): 418.

26 R.P., Siquijor, *Capital Development Plan*, no pagination; and United States, Census Office of the Philippines Islands, *Census of the Philippines Islands 1918*, vol. 3, *Agriculture, Medicinal Plants, Forest Lands and Proper Diet* (Manila: Bureau of Printing, 1920), pp. 18-20.

TABLE 1. Changes in population, carabao, cattle herd size, and selected economically important export crops, 1903-81, Province of Siquijor.

Year	Population	Cattle	Carabao	Abacá (mtons)	Copra (mtons)	Tobacco (mtons)
1903	50,156	368	8,174	-	-	-
1918	56,774	276	3,976	65.1	765.3	109.1
1939	59,507	7,763	6,597	4.7	2,135.8	96.4
1948	57,258	10,469	4,794	-	-	-
1960	59,555	13,436	3,903	6.8	-	.2
1970	62,976	15,028	1,924	.1	-	0.0
1981	70,360	16,670	1,110	0.0	4,639.0	0.0

SOURCES: U.S. Bureau of the Census, *Census of the Philippines Islands 1903*, vol. 4, *Agriculture, Social and Industrial Statistics*, (Washington, D.C.: Government Printing Office, 1905), pp. 229-237; U.S., Census Office of the Philippines Islands, *Census of the Philippines Islands 1918* (Manila: Bureau of Printing, 1920), vol. 1, *Geography, History, and Climatology*, p. 224; vol. 2, *Population and Mortality*, p. 243; vol. 3, *Agriculture, Medicinal Plants, Forest Lands and Proper Diet*, pp. 23, 365. Commonwealth of the Philippines, Commission of the Census, *Census of the Philippines: 1939*, vol. 3, *Reports by Provinces for the Census of Agriculture* (Manila: Bureau of Printing, 1940), pp. 1277, 1281. Republic of the Philippines, Department of Commerce and Industry, Bureau of Census and Statistics, *Census of the Philippines: 1948*, vol. 2, pt. 3, *Report by Province for Census of Agriculture* (Manila: Bureau of Printing, 1953), pp. 1289-1292. Idem, *Census of the Philippines 1960*, vol. 1, *Agriculture Report by Province* (Manila: Bureau of Printing, 1961), pp. 37-21. Republic of the Philippines, National Economic and Development Authority, National Census and Statistics Office, *1971 Census of Agriculture: Negros Oriental*, vol. 1, *Final Report* (Manila: Bureau of Printing, 1971), pp. 16-30. Idem, *1980 Census of Agriculture: Siquijor*, vol. 1, *Final Report* (Manila: Bureau of Printing, 1980), pp. 6-9.

Underscoring the importance of the island's agricultural-based economy are the 1980 census figures, which indicate that an estimated 77 percent of all occupations derive from natural resource sectors consisting of agriculture, animal husbandry, forestry, fishing, and hunting.[27] However, commercial forest and fishery production is minuscule; rattan, nipa shingles, orchids, and *salago* bark for use in making handicrafts dominate the exports in the former and small quantities of algae, seashells, and fish fry dominate the latter. Hunting is all but nonexistent. As a result, agricultural production (including animal husbandry) dominates both the local and export sectors, supporting an estimated farming population of 8,500. Some 60 percent of the insular area is under some form of agricultural production, led by coconut (20 percent), corn (18 percent), and cassava (6 percent). An additional 39 percent is in pasture or other open land).[28]

---

27 R.P., National Economic and Development Authority, National Census and Statistics Office, *1980 Census of Agriculture: Siquijor*, vol. 1, *Final Report* (Manila: National Census and Statistics Office, 1980), p. xxviii.

28 R.P., Siquijor, *Capital Development Plan*, no pagination.

Chief among non–commodity-based economic sectors are off-island remittances and revenue derived from manganese and guano mining operations.[29]

In sum, the island's exports are dominated by copra and livestock, and imports are dominated by grains, animal feeds, construction materials, metals, and petroleum products.

Of particular relevance to the present paper's theme is the evolution of the cattle industry's role in the provincial economy. The Spanish initially emphasized cattle production. Although carabao preceded the Spanish in the Philippines, the Spanish were instrumental in bringing the first cattle to the islands, in all likelihood breeds originating from China and Nueva Espana. These breeds were small and poorly adapted to the local conditions, and a viable cattle industry failed to become established.[30]

Few additional efforts at breeding appear to have been attempted until the onset of the American period and the creation of the Bureau of Agriculture in 1902. Findings of a cattle census conducted by the Americans for purposes of assessing supplies of meat for the army indicated that the country was an extensive importer of livestock and meat products with no available commercial supply. The absence of sufficient quantities of local livestock (particularly in the presence of several cattle diseases, of which rinderpest was the most serious) was a major reason why the army began the importation of meat supplies, chiefly from Australia, and, at least in part, was the justification for the establishment of the new bureau.[31] Since then cattle production has more than quadrupled, and livestock-rearing has evolved into a significant national industry.

At the provincial level the industry's development seems to have followed much the same pattern. At the turn of the century cattle appeared to have had little significance in the local economy, totaling only 368 head in 1903 (table 1).[32] However, since that time livestock production has grown to a family farm average exceeding one cow per provincial household and an insular herd size estimated to be 16,670 in 1981, accounting for some 6.6

---

29 R.P., Bureau of Mines, *Preliminary Report*, p. 19.

30 One effort dating back to the sixteenth century entailed introducing some 24,000 head of cattle in the islands, but physical constraints including heat and absence of pasture prevented further increases in herd size. Zoilo M. Galang, *Encyclopedia of the Philippines*, vols. 5 and 6, *Commerce and Industry* (Manila: Exequiel Floro, 1950), p. 10.

31 G.E. Nesom, "The Practicability of Supplying Native Beef to the Army," *Philippine Agricultural Review* 6 (March 1911): 121; and Tate, *Modern South-East Asia*, p. 444.

32 This conclusion is supported by the absence of cattle in exports listed in 1908. United States, War Department, Bureau of Insular Affairs, *A Pronouncing Gazetteer and Geographical Dictionary of the Philippine Islands* (Washington, D.C.: Government Printing Office, 1902), p. 370; and United States, War Department, *Report of the Philippine Commission. Part 1, 1908* (Washington, D.C.: Government Printing Office, 1908), p. 396.

percent of the region's total cattle population and contributing a significant share to family income.[33]

Despite the overall trend, absolute herd size appears to have fluctuated significantly with the vagaries of nature and people. In the late nineteenth century an epidemic of rinderpest resulted in the destruction of an estimated 75 percent of the national herd.[34] While no figures were obtained from Siquijor prior to 1903, an early census from Bohol (of which Siquijor was then a subprovince) showed a decline from 18,389 to 15,149 in the years between 1881 and 1903.[35] During and immediately following World War II, excess cattle in the national herd were slaughtered for food, contributing to a decrease in herd size estimated at 50 to 66 percent.[36] Again, while no figures are available for Siquijor during this period, data from the Province of Negros Oriental (of which Siquijor had become a subprovince) showed a drastic decline from 59,564 to 6,290 between the years 1938 to 1945.[37]

In addition to growing market demand, there may have been other factors contributing to increases in the provincial herd size. Early censuses indicated that the carabao dominated the provincial herd for use as work animals. Although carabao continue to be used as a traditional source of power where water is available, there has been a growing trend toward replacing these animals with cattle in areas where water is scarce (for instance, in the sugar regions such as Negros). Past government policies may also have influenced farmers' preferences toward the two animals. Traditionally

---

33 In a study from the period 1976 to 1978 some 562 head, or 74 percent, of the cattle slaughtered from the Municipality of Siquijor were slaughtered for market. As a source of family income a 1975 census estimated that over 35 percent of the households derived their major source of livelihood from raising livestock and poultry, with farming and gardening making up an additional 30 percent. See Galang, *Encyclopedia*, pp. 6-54; Finina N. Valencia, "Survey of the Status of Small Scale Backyard Livestock Raising in Siquijor" (B.A. thesis, Foundation University, 1979), p. 9; and the Commonwealth of the Philippines, Commission of the Census, *Census of the Philippines: 1939*, vol. 3, *Reports by Provinces for the Census of Agriculture* (Manila: Bureau of Printing, 1940), pp. 1266-1268.

34 Director of Agriculture, "The Animal Disease Problem," *Philippine Agricultural Review* 1 (May 1908): 190.

35 United States, Bureau of the Census, *Census of the Philippine Islands 1903*, vol. 4: *Agriculture, Social and Industrial Statistics* (Washington, D.C.: Government Printing Office, 1905), p. 235.

36 See Jovencio M. Bacalso, "The Cattle and Carabaos in Los Banos, Calamba and Cabuyao, Laguna after the Liberation of the Philippines in 1945," *Philippines Agriculturalist* 35 (March 1951): 163; W.J.A. Payne, "The Role of the Cattle Industry in the Philippines," *Philippines Journal of Animal Science* 3 (December 1966): 14; United States, War Department, Bureau of Insular Affairs, *Reports of the Philippine Commission* (Washington, D.C.: Government Printing Office, 1904), p. 357; and J.E. Spencer, *Land and People in the Philippines* (Berkeley: University of California Press, 1952), p. 174.

37 Republic of the Philippines, Department of Agriculture and Natural Resources, *Soil Survey*, p. 28.

carabao have been protected from slaughter for meat unless the animals were considered unfit for draft purposes. Original legislation extends back to the rinderpest epidemics in the national herd in 1878-92. People began slaughtering their carabaos and selling the meat in the belief that the disease would otherwise spread to their unaffected animals, rendering them unsalable. More recent presidential decrees have reinforced the policy.[38] As a result many people came to believe that cattle were superior to carabao because of both their greater tolerance to climatic conditions and their flexibility in use (and disposal) unfettered by government regulation, which in all likelihood contributed to the precipitous decline of carabao in the province (table 1). In light of Siquijor's small farmer economy these factors seem largely to explain the growth in the provincial herd. In this marginal economy of smallholders, cattle raising occurs alongside other farming activities, and highly adaptable, useful, and easily disposable animals providing an additional source of income represent an optimal use of scarce resources.

Grazing is dominated by open pasturing on both public and private lands and is often complemented by cut-and-carry methods, especially in the dry season. It is the traditional open grazing of cattle, however, that appears to be a major factor contributing to the erosion problem. Evidence of land degradation associated with this practice has been documented in the Philippines as early as the mid-1930s. However, this was attributable not only to pressures associated with increasing cattle populations, but to growing competition for open pasturelands for alternative agricultural uses.[39]

Whereas growth in herd size and grazing practices appear to be important factors explaining the widespread erosion in Siquijor, another trend may have exacerbated the situation, namely land fragmentation. However, before examining its role, a better understanding of the island's recent population and demographic trends is required.

Between 1903 and 1980, the year of the last census, Siquijor's population grew from 50,156 to 70,360 (table 1). The latest figures signify Siquijor's rank as the country's third smallest province in terms of population, but place it in the top twenty nationally and second regionally (after Cebu) with respect to density (240 people/km^2), but like the Philippines as a whole (and many other countries in Asia), Siquijor's population is largely rural, with only an estimated 11 percent classified as urban.[40]

---

38 David C. Kretzer, "How to Build Up and Improve a Herd or Flock," *Philippine Agricultural Review* 20 (July 1928): 220-221.

39 Miguel Manresa, Pepito N. Narciso, and Abel L. Silva, "Comparative Efficiency of Pasture Management Methods," *Philippine Agriculturalist* 27 (October 1938): 343.

40 R.P., Siquijor, *Capital Development Plan,* no pagination; and Republic of the Philippines, National Economic and Development Authority, *1980 Census*, p. xxi.

The annual growth rate was 1.4 percent per year for the period 1948 to 1980—far below the 3 percent recorded for the nation as a whole over this same period. Lagging population growth is largely attributable to a combination of decreases in live births and out-migration, the latter typical of a larger trend in the Central Visayas characterized by large demographic shifts to metropolitan Manila and Mindanao.[41]

Emigration has long played a significant role in both the region generally and Siquijor specifically. At the beginning of the twentieth century growing demand for labor in Hawaii's expanding sugar industry created new opportunities for the un- and under-employed. Continuing problems with recruiting Chinese and Japanese labor together with the recent annexation of the Philippines combined to create a market for cheap Filipino labor. In the period 1909-29 approximately 106,000 Filipinos emigrated to the Hawaiian Islands. Although the Ilocos region (Luzon) dominated the population source areas, islands of the Visayas ranked a close second in supplying labor. The aggregated data prevent a specific accounting of Siquijor's contribution to this total, but it is highly probable that as part of the Central Visayas it was significant in proportion to the island's population. The Central Visayan provinces were identified as labor-surplus areas by the Philippine government and thus made accessible to the Hawaiian Sugar Planters Association (HSPA), the prime recruiter at the time, justifying the establishment of one of their two national recruiting offices in Cebu. Of the total 74,000 recorded out-migrants between the years 1916 and 1928 some 20,000 or 27 percent originated from the Central Visayan provinces.[42]

Out-migration has turned increasingly domestic over the period 1948 to 1970. This has been the result of the desire to leave crowded, resource-poor lands characteristic of the region for the less-developed and still relatively resource-rich islands typical of the southern Philippines.[43] Migrants

---

41 Siquijor's rate of out-migration was estimated at 15 per 1,000 for the period 1975-80, or 1.5 percent. Republic of the Philippines, National Economic and Development Authority, *1980 Census*, p. xxi; and idem, *Central Visayas*, p. 10.

42 See Mary Dorita, "Filipino Immigration to Hawaii" (M.A. thesis, University of Hawaii, 1954) for a detailed account of the evolution of Hawaiian immigration, and Bruno Lasker, *Filipino Immigration to Continental United States and to Hawaii* (Chicago: University of Chicago Press, 1931) for broader and more introspective accounts of the subject.

43 For example between 1960 and 1970 the population of Mindanao increased almost three times as fast as that of the Central Visayas (48 vs. 18 percent). See Republic of the Philippines, Department of Commerce and Industry, Bureau of the Census and Statistics, *Net Internal Migration in the Philippines, 1960-1970*, by Yun Kim, Technical Paper no. 2 (Manila: Bureau of the Census and Statistics, 1972), p. 25; and Republic of the Philippines, National Economic and Development Authority, National Census and Statistics Office, *Social, Economic and Demographic Factors Relating to Interregional Migration Streams in the Philippines: 1960-1970*, by Oscar F. Palmeras, UNFPA-NCSO Population Research Project, Monograph 11 (Manila: National Census and Statistics Office, 1977), p. 3.

appear to be chiefly agricultural workers and fishermen, a result of the region's poorly developed agricultural sector.

Unlike many areas of the Philippines, in Siquijor most farmers are fortunate enough to own and cultivate their own land, and tenancy does not appear to be significant. Average farm size is .96 ha (table 2); few holdings are larger than 10 ha (table 3) and none greater than 20 ha. Owing to differing definitions as to what constitutes a farm in the country's various censuses, the usefulness of other data comparisons is dubious.[44]

TABLE 2. Number, average size, and ownership patterns of farms, 1918-80, Siquijor.

Year	Number of farms	Average size (ha)	Number of owners[a]	Number of other tenants[b]
1918	11,913	1.02	9,199	2,714
1939	7,827[c]	1.62	6,600	1,227
1948	7,598[c]	1.43	-	-
1960	7,690[c]	1.70	6,024	1,666
1971	6,653[d]	1.32	5,739	909
1980	10,024[e]	.96	8,326	1,698

SOURCES: U.S., Census Office of the Philippines Islands, *Census of the Philippines Islands 1918*, vol. 3, *Agriculture, Medicinal Plants, Forest Lands and Proper Diet* (Manila: Bureau of Printing, 1920), pp. 82-83. Commonwealth of the Philippines, Commission of the Census, *Census of the Philippines: 1939*, vol. 3, *Reports by Provinces for the Census of Agriculture* (Manila: Bureau of Printing, 1940), p. 1262. R.P., Department of Commerce and Industry, Bureau of Census and Statistics, *Census of the Philippines: 1948*, vol. 2, pt. 3, *Report by Province for Census of Agriculture* (Manila: Bureau of Printing, 1953), p. 1287. Idem, *Census of the Philippines 1960*, vol. 1, *Agriculture Report by Province* (Manila: Bureau of Printing, 1961), p. 37-3. R.P., National Economic and Development Authority, National Census and Statistics Office, *1971 Census of Agriculture: Negros Oriental*, vol. 1, *Final Report* (Manila: Bureau of Printing, 1971), pp. 1-2. Idem, *1980 Census of Agriculture: Siquijor*, vol. 1, *Final Report* (Manila: Bureau of Printing, 1980), p. 1.

[a] Includes full and part-owners.

[b] Includes cash, labor, and crop tenants.

[c] Farm defined as "parcel or parcels of land having a total area of at least 1,000 mt^2 which are used for the raising of crops . . . includes all parcels . . . actually worked by one person." Commonwealth of the Philippines, Commission of the Census, *Census of the Philippines: 1939*.

[d] Farm definition expanded to include "Any land . . . used for raising at least 20 heads of livestock and/or 100 heads of poultry." R.P., National Economic and Development Authority, National Census and Statistics Office, *1971 Census of Agriculture*.

[e] Farm definition expanded to include "10 hectares . . . under permanent meadows and/or pastures, raising 10 heads of large animals . . . 20 heads of small animals, 100 heads of poultry, 50 heads of rabbits." Idem, *1980 Census of Agriculture*.

---

44 For example, large increases in farm numbers for census year 1980 appear to be due to a broadening of the definition for farm from that used in earlier censuses.

TABLE 3. Changes in farm size, 1918-80, Province of Siquijor.

Size (ha)	1918	1939	1960	1971	1980
<1	7,988	3,416	2,117	2,728	6,549
1<2	2,576	2,755	3,083	-[a]	2,340
2<5	1,028	1,460	2,220	-[a]	976
5<10	268	177	213	104	142
>10	53	15	57	1	17
Total	11,913	7,827	7,690	6,653	10,024

SOURCES: U.S., Census Office of the Philippines Islands, *Census of the Philippines Islands 1918*, vol. 3, *Agriculture, Medicinal Plants, Forest Lands and Proper Diet* (Manila: Bureau of Printing, 1920), p. 110. Commonwealth of the Philippines, Commission of the Census, *Census of the Philippines: 1939*, vol. 3, *Reports by Provinces for the Census of Agriculture* (Manila: Bureau of Printing, 1940), p. 1265. Republic of the Philippines, Department of Commerce and Industry, Bureau of Census and Statistics, *Census of the Philippines 1960*, vol. 1, *Agriculture Report by Province* (Manila: Bureau of Printing, 1961), p. 37-4. Republic of the Philippines, National Economic and Development Authority, National Census and Statistics Office, *1971 Census of Agriculture: Negros Oriental*, vol. 1, *Final Report* (Manila: Bureau of Printing, 1971), p. 4. National Economic and Development Authority, National Census and Statistics Office, *1980 Census of Agriculture: Siquijor*, vol. 1, *Final Report* (Manila: Bureau of Printing, 1980), p. 1.

[a] Figures were not disaggregated for these size classes.

However, these figures may not be representative of the island's upland areas. One study of selected Central Visayan upland communities found Siquijor's average farm size and population density per farm hectare to be .36 ha and 14.7 respectively, the smallest and highest values among the seven upland study sites. Of particular significance was the finding that Siquijor led all sites in cattle density, with an average count of 4.3 per farm for farms where cattle were present (42 percent).[45] That these characteristics and trends significantly affect demands on the land and in turn affect erosion rates is readily apparent.

---

[45] The study also described an intensification of production higher in Siquijor than in any other site, characterized by the use of a near-universal third cropping season for corn, an above-average percentage of landownership (44 percent), land continuously in production (estimated between 99.4 and 99.7 percent), use of fertilizers and other soil amendments, absolute tenancy (31 percent), local consumption of agricultural production, but a lower percentage of farm parcels (2.5 per farm) and soil fertility and yields. One explanation for these characteristics suggests that declines in farm size due to inheritance have resulted in a growing number of uneconomical enterprises forcing some farmers to sell, in turn resulting in consolidation of landholdings. See Lionel Deang et al. "Economic Life in Seven Upland Areas of Region VII," paper prepared for the University of San Carlos Office of Population Studies, Cebu City, 1981; and Lionel Deang, Josephine L. Avila, and Froilan Lirasan, "Economic Life in Seven Upland Areas of Region VII: 1981," *University of San Carlos Research Digest Series* 12 (December 1985): 4-5.

Institutional Framework

The Philippines is fortunate to have a large body of legislation and administrative machinery that addresses a range of environmental issues.[46] National legislation affecting land use potentially addressing the erosion/ sediment conflict can be classified into one of three types. The first is based on the formulation of national guidelines and standards to be used by local planning authorities to develop land-use plans. One function of the Ministry of Human Settlements, established in 1978 but now defunct, was to develop standards and guidelines to assist local governments in developing land-use plans and zoning ordinances. It was the intention that environmental criteria related to air and water quality, land use, and waste management would be incorporated into these planning and management tools for purposes of assessment and modification, if necessary, of newly proposed development projects and programs.

A second approach has been to empower various government sectors to designate selected areas as critical or in some other way unique, providing the legal basis for government intervention to modify land use.[47] The power to designate such areas was expanded under the Filipino Environmental Impact System, which gave the president the right to declare areas environmentally critical. This in turn prevented development in the absence of an environmental compliance certificate issued by the president's office.[48]

A third approach has been to codify specific uses of public lands based on previously determined criteria.[49]

A less expansive approach is characterized by the establishment of legal authorities to manage or promote the future development of designated

---

46 This includes a national environmental policy (PD 1151), environmental code (PD 1152), and an environmental impact statement procedure along the lines employed in the United States (PD 1586). See Republic of the Philippines, "The Philippine Environmental Impact System," *Supplement to Official Gazette* 78, no. 25, June 21, 1982.

47 For example the Environmental Code empowered the Human Settlements Commission, in coordination with other agencies, to formulate land-use schemes that would include "a method for exercising control by the appropriate . . . agencies over the use of land in areas of critical environmental concern." Moreover the policy toward soil conservation states that "the national government . . . shall . . . undertake a soil conservation program including . . . the identification and protection of critical watershed areas, encouragement of scientific farming techniques, physical and biological means of soil conservation . . . for effective soil conservation." See Republic of the Philippines, National Environmental Protection Council, *Philippine Environmental Laws* (Quezon City: National Environmental Protection Council, 1981), pp. 37-40.

48 Ibid., p. 54.

49 For example the Forestry Code defines agricultural lands as those areas with slopes less than 18 percent. Lands with slopes in excess of 18 percent are classified as public lands and must be kept in permanent forest cover.

areas with implicit if not explicit objectives to mitigate present and possible future environmental disturbances. These authorities' mandates are usually highly limited and are often advisory or coordinative in nature.[50]

Other relevant legislation can be found under the Filipino Water Code. Existing water law observes riparian rights and specifically protects the rights of the lower riparian.[51] Of special note is the right to declare "any watershed or any areas of land adjacent to any surface water . . . by the Department of Natural Resources as a protected area . . . [r]ules and regulations may be promulgated . . . to prohibit or control . . . activities by the owners . . . within the protected area which may damage or cause the deterioration of the surface water."[52]

There also exists legislation addressing the dumping of wastes into water bodies, though much of this is oriented to industrial pollution. However, Commonwealth Act 383 prohibits the dumping of refuse or substances of any kind into any river which could bring about the rise or filling of riverbeds or cause artificial alluvial formations.[53]

With respect to coastal resource conflicts with upland source areas, there appears to be no extant legal machinery useful for conflict mitigation. Because some fifty government institutions are responsible for one or more activities in the coastal area, an interagency task force was set up to coordinate activities among the various political entities. Before falling into disuse, the task force was responsible for a number of activities, including compiling a coastal resources inventory, sponsoring a study of existing land uses, and initiating a few pilot activities. Nevertheless, there is no evidence that the task force addressed in any significant manner issues resulting from upland/coastal conflicts, particularly with respect to small-scale rural land use.[54] However, for purposes of promoting local economic development

---

50 These include the Pasig River Development Council, the Bicol River Basin Program, and the Laguna Lake Development Authority (LLDA). An exception to the generalization regarding mandates is the LLDA, which is empowered to "pass over all plans, programs, and project developments proposed by local government agencies within the region . . . related to development of the region [and] . . . determine . . . if . . . developments need to be approved by the Authority." See Republic of the Philippines, National Environmental Protection Council, *Philippine Environmental Laws*, p. 162.

51 Under article 50 of the Water Code the lower riparian is legally required "to receive the waters which naturally . . . flow from higher estates as well as the stone of earth which they carry with them." See Republic of the Philippines, National Water Resources Council, *Philippines Water Code and the Implementing Rules and Regulations* (Quezon City: National Water Resources Council, 1982), p. 13.

52 See ibid., article 67, p. 17.

53 Ibid., p. 461.

54 Amado S. Tolentino, Jr., "Philippine Coastal Zone Management: Organizational Linkages and Interconnections," *Environmental and Policy Institute Working Paper* (Honolulu: East-West Center, 1983).

Siquijor was declared a tourist zone in November 1978, which conceivably could provide a basis for regulating upland land-use activities affecting coastal resources. Presidential Decree 1801 provides penalties for enterprises threatening the degradation of resources considered of value for tourism promotion. In Siquijor these include the island's beaches and reefs.[55]

In addition to national legislation, a substantial amount of the responsibility for environmental issues appears to fall into the preserve of local government. This has been a relatively recent development. From Spanish times to the present the Philippines has always had a centralized system of government, with minimal delegation of authority to the provinces, and local authorities in effect were the agents of their national counterpart. The country's administrative units are the regions, provinces, municipalities, and *barangays* (communities). In addition there is the designation of city. Each unit is empowered to execute its appropriate resolutions and ordinances as long as no conflict occurs with the next higher unit's ordinances or with national law. Each unit has the dual responsibility of serving as agent of the national government and providing local government services. Despite recent attempts at devolution of power to the regions little appears to have changed in control from the center, particularly with regard to taxing authority, and most services are administered by the national government through regional field agencies.[56] However, local responsibility for environmental matters has been explicitly provided for in the law.[57] In Siquijor there have been no fewer than seven resolutions promulgated since 1972 (the year the island became a full province) that address limits on cattle exports, all directed at conserving local supplies and guarding against rises in consumer meat costs.[58]

---

[55] Republic of the Philippines, Presidential Proclamation, "Declaring Certain Islands, Coves, and Peninsulas in the Philippines as Tourist Zones and Marine Reserves under the Administration and Control of the Philippine Tourism Authority," no. 1801 (Manila, Philippines Tourism Authority, November 10, 1978); and Presidential Decree, "Revising the Charter of the Philippine Tourism Authority created under Presidential Decree no. 180, Dated May 11, 1973," no. 564 (Manila, Philippines Tourism Authority, October 2, 1974).

[56] Russell J. Cheetham and Edward K. Hawkins, *The Philippines: Priorities and Prospects for Development* (Washington, D.C.: World Bank 1976), p. 390; and G. Ayson and J.P. Abletez, *Barangay: Its Operations and Organizations* (Manila: National Book Store, 1987), p. 15.

[57] Section 58 of the Environmental Code states that local governments should actively participate and support environmental management programs of the government. Further, laws have given local government officials the right to enforce a range of national and local environmental laws, including prohibiting "the denudation of forest or watershed that precipitates soil erosion." See Republic of the Philippines, National Environmental Protection Council, *Philippine Environmental Laws*, pp. 46-51.

[58] See Siquijor, Office of the Governor, resolutions 47 (June 1, 1973); 55 (June 15, 1973); 63 (July 12, 1973); 74 (November 22, 1974); 58 (November 4, 1975); 14 (February 15, 1978); and 81 (July 18, 1984).

At the level of municipality the Maria Municipal Council has prohibited the pasturing of large animals (defined as cattle, carabao, sheep, goats, and horses) on private lands in the absence of owner consent.[59] There is also an ordinance calling for setback zones from municipal waters (which include rivers) varying in distance according to adjacent land use.[60]

Regarding coastal waters the council has declared a zone extending three miles offshore from the low-tide mark and bounded by two perpendicular lines extending from the municipality's terrestrial boundaries to be municipal waters. Ordinances governing these waters regulate fishing, designate special-use areas (e.g., for mangrove replanting and placing of artificial reefs), and establish a marine park.[61]

A survey of resolutions taken by *barangay* councils revealed no relevant resolutions.

---

59 The penalty for infringement is a fine of ten *pisos* (approximately fifty cents) and/or twenty days in prison. The procedure for enforcement is for the plaintiff to take the animal to the nearest office of the Integrated National Police and file a complaint. To collect the animal the owner must pay the fine. See Siquijor, Office of the Municipal Council, resolution 2 (July 3, 1977).

60 See ibid., resolution 5 (June 14, 1974).

61 The council declared a marine reserve in January 1983 by which fish and coral reefs were protected within a designated zone in Maria Bay. However, no provision was made for regulating upland uses contributing to reserve degradation. See ibid., resolutions 16 (January 20, 1980); and 6 (January 28, 1987); and Republic of the Philippines, Siquijor, *Capital Development Plan*, no pagination.

Chapter 4

METHODOLOGY

Several objectives form the focus of this study: (1) to determine the nature and significance of increased rates of upstream erosion and sedimentation on coastal resources uses (the conflict); (2) to compare the respective perceptions of upland and coastal resource users with respect to the land-use practices and processes contributing to the conflict; (3) to identify the adjustments of coastal resource users in response to changing environmental conditions associated with increased rates of sedimentation; (4) to identify the major underlying social and economic forces contributing to land-use practices responsible for accelerated soil loss; and (5) to examine whether existing policies exacerbate or mitigate the conflict.

The Central Visayas Regional Project (CVRP) in the Philippines, a joint effort between the Government of the Philippines and the World Bank, was designed to improve the living standard of small-scale farmers, fishermen, and forest dwellers in that region (figure 1). It is being implemented over a five-year period (1984-89) at five sites in each of the region's four provinces (Cebu, Bohol, Siquijor, and Negros Oriental). Sites were selected that met provincial development priorities and exhibited severe environmental degradation, widespread human impoverishment, and potential development opportunities.[1]

The Maria watershed on the island province of Siquijor (figure 2) was selected for this study because of (1) its small size and short drainage length, which minimized the need to account for other intrawatershed activities affecting coastal processes and resources; (2) existing nonsustainable land-use practices in its upper reaches, which contribute to a discernible modification of watershed processes; and (3) conditions representative of other areas.

The research design was based on a hypothetical model developed to explain existing conditions observed in visits to the site prior to initiation of

---

1 Associacion ng mga Consultants na Independente (Philippines) Inc., "Central Visayas Regional Project-1: Mid-Project Review Report," Cebu City, 1986, p. 2 (typewritten).

the data collection effort. The model suggested that overpopulation and a limited natural resource base has caused encroachment and nonsustainable exploitation of marginal upland areas typified by intense overgrazing. This is contributing to accelerated soil erosion which in turn is resulting in one or more downstream impacts characterized by water-borne sediment affecting existing lowland and nearshore resource-user systems. The degree to which coastal impacts are associated with upland land uses and processes is unknown, but is predicted to be correlated with subsequent adoption of adjustments among coastal inhabitants in response to changing environmental conditions associated with the conflict. In addition, a number of factors may be exacerbating the situation. These include land fragmentation, absentee ownership, and variations in tenure system. Finally, existing policies (or their absence) appear to be contributing to these upland pressures.

The research format selected was the case study,[2] divided into two discrete but interrelated components: a biophysical assessment and a socioeconomic study. Owing to the evidence of severe overgrazing at the study site, upland research elements under the two components focused on this form of land use and the underlying social dynamics contributing to it.[3]

The Biophysical Component

A key constraint to research in the tropics is the absence of baseline data collected over a sufficient time period to prove useful for comparative purposes. This issue is particularly relevant to the present study's biophysical component, where high variability in meteorological phenomena (such as typhoons and monsoonal storms) can greatly affect erosion rates and sediment yields. To address this constraint a resource assessment is commonly used for purposes of identifying priority problems, degree of significance, etc. In the present study it has been used to provide a picture of existing conditions and serve as a basis from which to compare results of the second and third components. Findings from the assessment were compared with re-

---

2 The value of case studies in research design has been much debated. Criticism stems primarily from the absence of baseline information and controls. Contrasting views hold that case studies are beneficial in providing the investigator a means to explore a problem when the significant variables and relationships of the phenomenon in question are not well known. To address the former criticism a significant level of effort was allocated both to a detailed biophysical assessment of the river basin and to a historical review to complement field studies. See D.T. Campbell and J.C. Stanley, *Experimental and Quasi-Experimental Designs for Research* (Chicago: Rand McNally, 1966); and L.A. Salter, *A Critical Review of Research in Land Economics* (Madison: University of Wisconsin Press, 1967) for contrasting views on the validity of case study as a research design.

3 The evidence observed for severe overgrazing included uneven grazing pressures, shortened grass leaves and stems, highly compacted soils, numerous landslips, and an abundance of cattle.

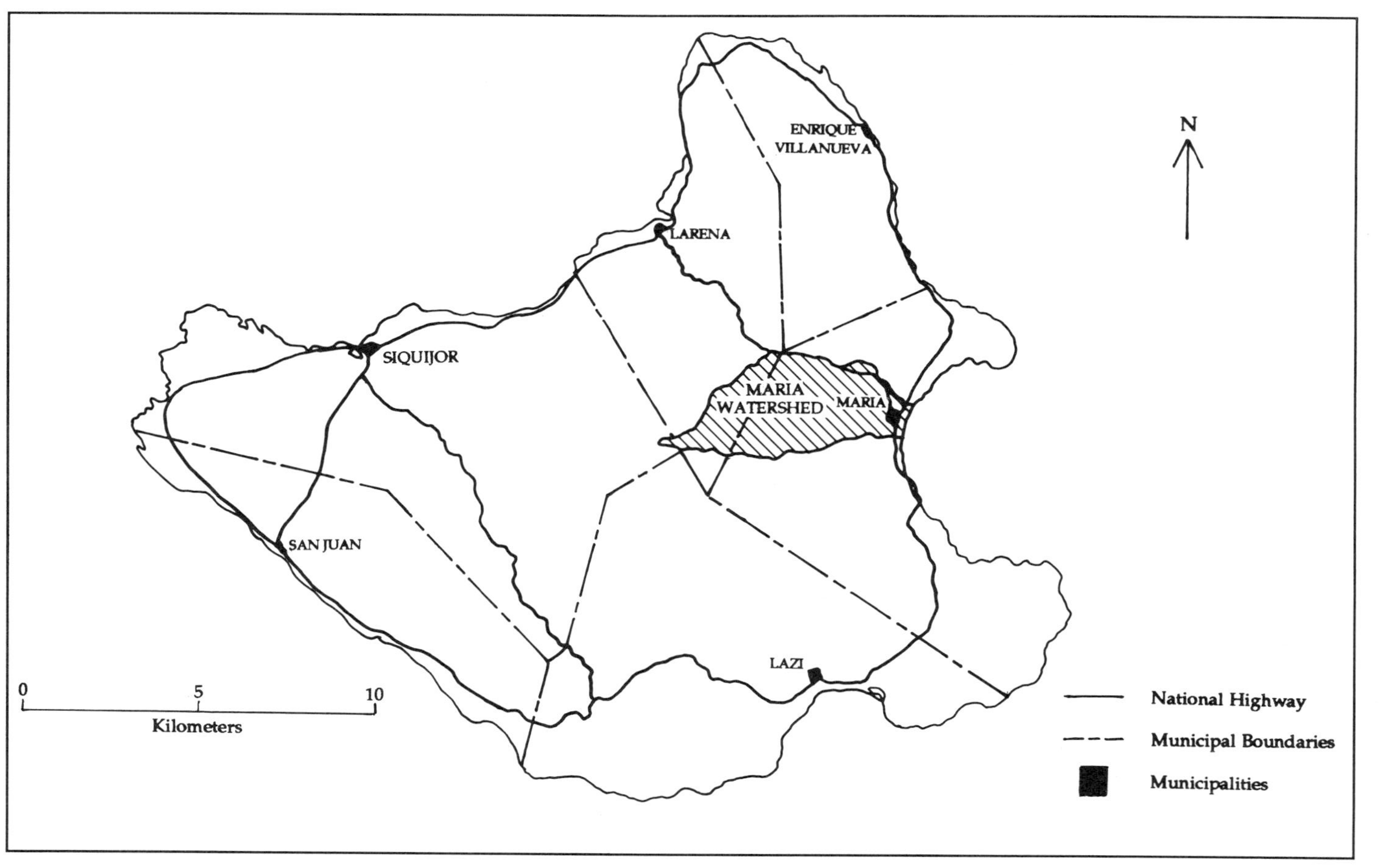

Fig. 2. Province of Siquijor

sponses from local inhabitants derived from discussions and a survey questionnaire. Finally, there exist certain processes relevant to the study at hand that are cyclical rather than episodic in nature (such as the coastal current regime), which were capable of being adequately characterized within the study's time constraints.

The upland portion of the biophysical component consisted of monitoring rainfall, river flow, and suspended-sediment load. In addition erosion severity was estimated at randomly selected upland farms within the watershed.

Rainfall was measured from a network of rain gauges established at four stations varying in altitude and located along the periphery of the watershed.

River volume transport flow was calculated from measurements taken twice daily from a relatively straight 20 m section of the river. Flow velocity was measured through repeated timing trials of a uniformly shaped floating marker placed initially along the left, middle, and right bank sections.[4] Suspended-sediment load was determined with use of a bottle sampler.[5]

Data collection for storm flow heights and sediment load was facilitated by the placement of a staff gauge in the river. During rainfall, readings of gauge height and sediment samples were taken on an average of every fifteen minutes until water levels returned to preflood levels. Rating curves were established from daily monitoring and storm observations over the course of a year that correlated staff height with measured flow and suspended sediment respectively. These data in turn were used to calculate mean annual suspended-sediment discharge for the study period, using the flow-duration sediment rating curve method.[6]

Severity of erosion was determined for farms selected at random from a sampling frame developed for a survey questionnaire. Samples were taken by a random toss and the use of an auger. Measured depths for each soil type were compared to standard erosion category levels used by the Philippine Department of Agriculture's Bureau of Soils to ascertain erosion severity.

---

4 These are straightforward procedures and are discussed at length in Thomas Dunne and Luna B. Leopold, *Water in Environmental Planning* (New York: W.H. Freeman and Co., 1978), pp. 655-656.

5 A hand-held bottle was used for collecting water near the bottom of the stream at mid-channel. The procurement of an integrated sampler was not possible within the study's time constraints. Samples were filtered using 45μ filters, oven dried, and weighed.

6 See Carl R. Miller, *Analysis of Flow-Duration Sediment-Rating Curve Method of Computing Sediment Yield* (Washington, D.C.: U.S. Bureau of Reclamation, 1951), pp. 1-15.

The coastal portion of this component consisted of monitoring the bay's current, rates of sedimentation, and selected water quality parameters. In addition estimates of living benthic cover and fish abundance and diversity were made at selected sites. The current regime was determined through the tracking of window-shade drogues in the water column.[7] Positions were taken from a boat and recorded together with time through the use of a hand-bearing compass and watch. Studies were initiated from a site in front of the river mouth calculated to be approximately 12 m in depth (figure 3).[8]

Sediment traps were placed at five offshore sites located in front of and on each side of the Maria River mouth (figure 3). Stations were evenly distributed alongshore at .5 km intervals in depths approximating 12 m.[9] Samples were collected on a monthly basis and analyzed for total dry weight. Bottom samples were also taken at each station and analyzed for grain size and composition.

Two water quality parameters, salinity and visibility, were selected for daily monitoring at each of the five sediment trap stations, owing to the great influence that terrestrial runoff can exert on their values. Salinity and visibility were measured with an optical refractometer and secchi disc, respectively.

Results of an initial survey indicated that small coral patch reefs and marine grassbeds were the marine communities most vulnerable to the effects of sedimentation in Maria Bay. These communities were systematically sampled by a transect and quadrat method. A total of twelve transects running perpendicular to the general linear orientation of the coast were completed on both sides of the river .25 km apart near the mouth but increasing to .5 km at greater distances. Samples were taken at the isobaths

---

7 A drogue consisted of a 1 m^2 piece of canvas with two parallel edges attached to a 1.5" diameter PVC pipe and steel reinforcing rod respectively. The PVC pipe was attached to a surface float by a 1 m length of polypropylene line. A floating wooden block and bicycle flag were also attached to the float to facilitate observations. A cold-chemical light stick was attached to the flag during night observations. For a detailed discussion on drogue design and utilization see Edward C. Monahan and Elizabeth A. Monahan, "Trends in Drogue Design," *Limnology and Oceanography* 18 (November 1973): 981-985.

8 Initially a total of eight studies were planned to characterize the tidal regime covering the ebb and flood phases of spring and neap tides for both monsoonal periods. However these were curtailed to six studies and limited with one exception to the southwest monsoon, owing to the severe sea conditions generated by the onset of the northeast monsoon and the scarcity of boats available to the project suitable for working in these conditions.

9 Traps consisted of a steel reinforcing bar attached to a .5 m diameter steel disk measuring 1/8" in thickness. Three holes were cut in the disk to place 6" diameter PVC pipes closed at one end. For a discussion of sediment trap technology and efficacy see W.B. Kirchner, "An Evaluation of Sediment Trap Methodology," *Limnology and Oceanography* 20 (July 1975): 657-660.

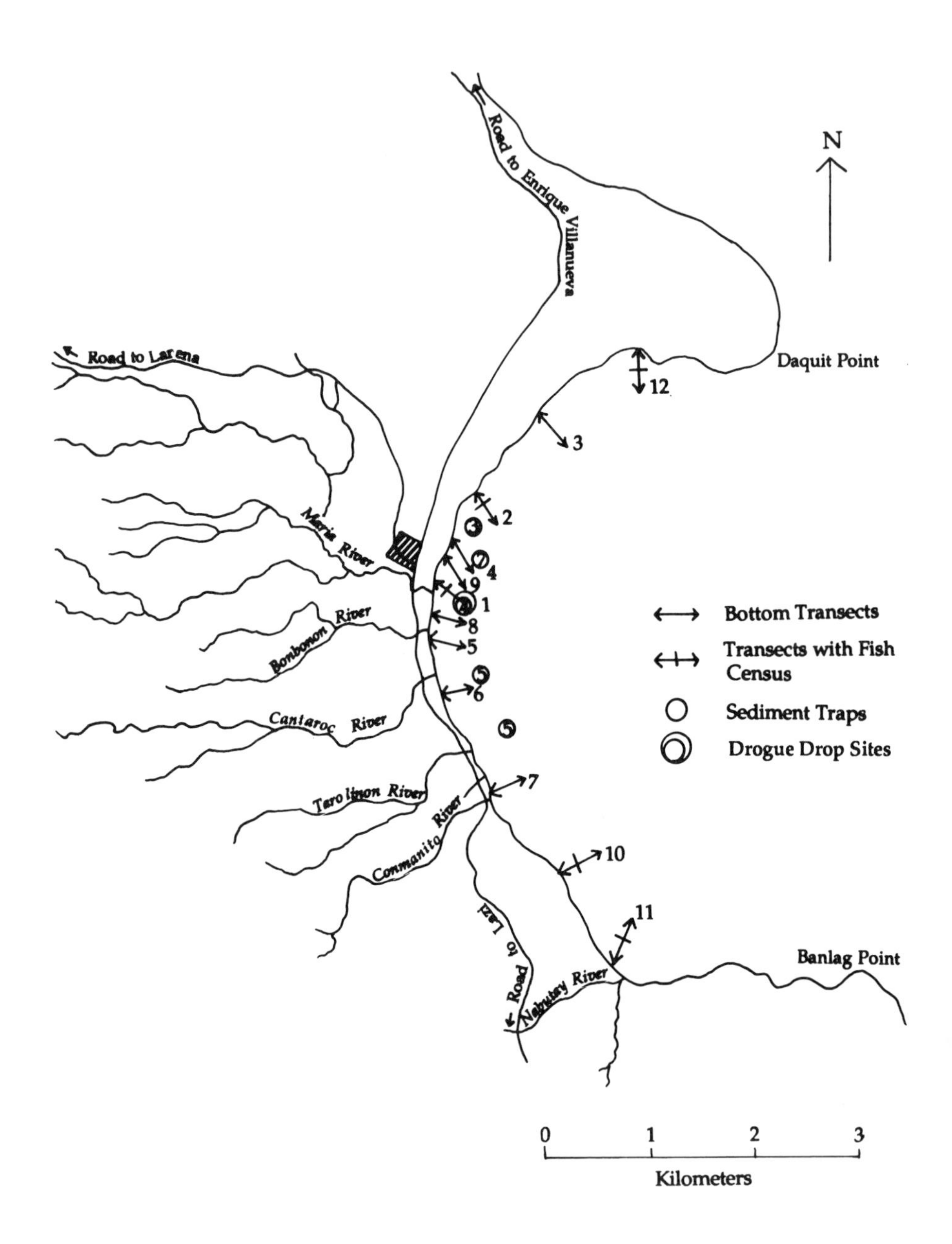

Fig. 3. Nearshore sampling sites, Maria Bay

3, 9, and 18 m along each transect. Predominant living and dead benthic cover was estimated for each sample in five 1 m quadrats, and sedimentation and its effects on the living communities were observed. Visual fish censuses were conducted at selected stations trimesterly to determine species diversity and biomass (figure 3).[10]

Data collection for both upland and coastal parameters began immediately following the advent of the rainy season (July 1987) and continued for a year.

Daily and periodic monitoring of rainfall, river flow, suspended sediment, and water quality was facilitated by the contractual hire and subsequent training of two farmers and one fisherman. Erosion severity sampling was conducted by personnel from the Bureau of Soils, Region 7, Cebu. Biotic and substrate sampling was conducted by scientists from the Marine Sciences Laboratory of Silliman University, Dumaguete. Finally, personnel from CVRP assisted in the studies of currents.

The Socioeconomic Component

Establishing the sources and magnitude of upstream-derived sediment and its significance for downstream users provides only a partial perspective on the conflict and processes in question. This aspect of the study attempted to: characterize the communities most likely to be affected by the erosion/sedimentation issue; explain potential differences in perception of erosion and sedimentation as hazards among the upland and coastal communities respectively; identify the range of adjustment to changing environmental conditions employed in downstream areas; examine the underlying social and economic forces contributing to existing land-use practices characteristic of Siquijor's upland farmers; and identify the effect of selected policies on the conflict.

The research was based primarily on a survey questionnaire complemented by two studies of the roles of land fragmentation and migration in determining land-use patterns at the study site.

For the survey's purposes the upland marginal lands farming sample was defined as farmers living in organized communities (this is the *barangay*, the smallest political unit in the Philippines) whose boundaries overlapped to varying degrees with the river basin's physical boundary and whose slopes equaled or exceeded 30° in at least 50 percent of the area. By these criteria six communities were identified and constituted the upland

---

10 See Garry Russ, "Distribution and Abundance of Herbivorous Grazing Fishes in the Central Great Barrier Reef. I. Levels of Variability across the Entire Continental Shelf," *Marine Ecology Progress Series* 20 (November 1984): 26-27, for a more detailed description of the methodology used.

sampling universe (Basac, Bonga, Candigum, Catamboan, Pisong B, and Upper Calunasan; figure 4).[11]

With regard to the coastal community three resource-user groups were identified as likely to be affected by downstream sedimentation associated with the upland farming of marginal lands. These were paddy farmers, coconut producers, and nearshore subsistence fishermen. A different approach was required in defining the sampling universe than used in the upland survey, for several reasons. First, owing to the river basin's constriction at its mouth only one community's boundaries overlapped with those of the river basin (Lo-oc), resulting in an extremely small population from which to develop a sample frame. Second, CVRP had been involved in education in the area and possibly exerted undue influence on community attitudes. As a result two adjacent coastal communities (Cantaroc A and B) were combined with Lo-oc and together represented the sampling universe (figure 4).[12]

Community captains were contacted and informed of the survey, which served to facilitate the compilation of a current list of heads of households. In the upland communities, a subset of heads of farming households was drawn from this list with the assistance of the captains and CVRP personnel and provided the basis for the sampling frame. Because of the high number of multiple occupations in the coastal community, livelihood classes were not discrete and no division by occupation was attempted. Sampling was random and limited to 102 and 120 respondents for the upland and coastal communities respectively.[13] Sampling was stratified in the three coastal communities and consisted of 40 respondents per community.

Two survey instruments were developed for the upland farming and coastal livelihood communities respectively (Appendixes 1 and 2). The two instruments shared a common set of core questions concerning basic personal information, recent weather characteristics, attitudes toward general well being, and responsibility of policies and institutions in contributing to or mitigating intrabasin resource-use conflicts.

---

11 *Barangay* Cang-apa was not surveyed, owing to the presence of the CVRP upland site management unit and the possible excessive influence it exerted on local attitudes, particularly with regard to awareness of poor land use and the need for sound land conservation measures.

12 Outside the Maria River basin, both communities shared similar upland physiography with Lo-oc. Attitudes to key questions that could be influenced by CVRP activities were compared between the communities to determine whether any statistical difference could be discerned.

13 The lower number was due to the difficulty in reaching farm lots in the rugged terrain typical of the river basin's uplands. The figure 102 represented a 25 percent sample of a total estimated farming family population of 410 for the six communities.

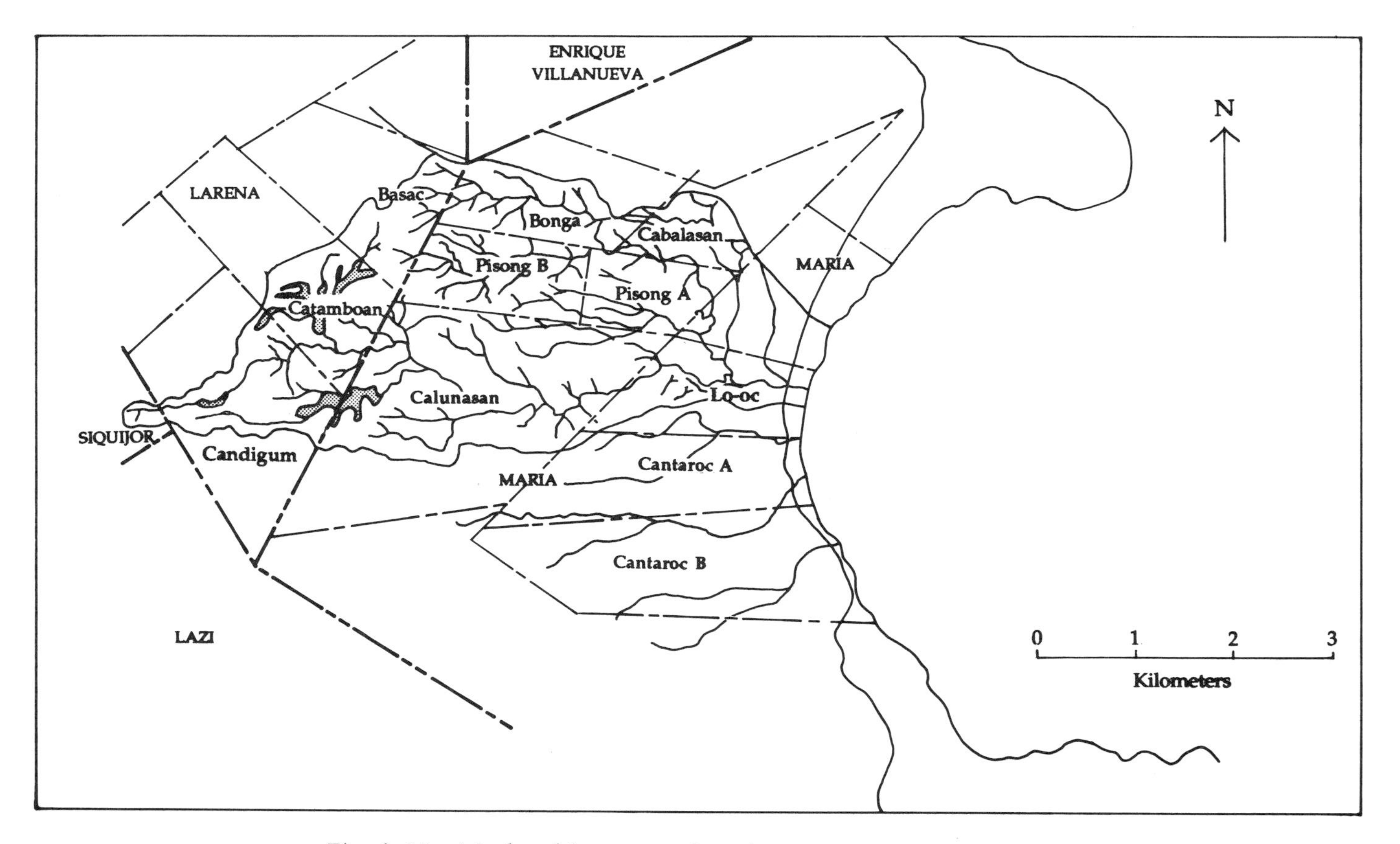

Fig. 4. Municipal and barangay *boundary map, Maria River basin*

In addition the upland community was questioned on specific livelihood characteristics, perceptions of erosion as a hazard, and to what extent they viewed the hazard's effect on other communities.

The coastal community was questioned about their livelihoods and primary source of income. The responses provided the basis for more detailed livelihood-specific questions concerning perceptions of sediment as an issue, its origins and eventual fate, and adjustments to the hazard.

Four types of questions were used in the two questionnaires: (1) Likert-type questions requesting the respondent to select one of seven responses ranging from "strongly disagree" to "strongly agree"; (2) open-ended questions; (3) yes/no questions; and (4) checklists. Differences in responses among inter- and intracommunity livelihood groups on the coast for Likert questions were measured statistically using a Kruskal-Wallis one-way ANOVA by ranks test. Differences in yes/no responses were tested for significance using a chi-square test at 1 and 5 percent significance levels.

Survey data from the coastal sampling frame were aggregated into livelihood groups when no significant differences were determined to exist between communities or the data did not meet the criteria required to apply one of the two tests identified above. Responses remained disaggregated when significant differences occurred within livelihood groups between communities and for certain questions addressing the effects of sediment on the quality of coastal natural resources and adjustments to same.

Prior to implementation of the questionnaires, a preliminary survey was conducted in a nonparticipating community for purposes of clarifying terminology and reducing levels of ambiguity. To overcome language constraints the survey was implemented by two social workers contracted from Silliman University.

In addition to the survey questionnaire, two underlying social issues, hypothesized in the model as being of potential significance in their relationship to land degradation, were examined in detail. The first concerned the relationship between population growth and land inheritance, its effect on land fragmentation, and in turn the role of fragmentation in influencing cattle:land grazing ratios over time. Study of the question was facilitated by the existence of landownership records in one community (Cang-apa) for the periods 1924-28 and 1983 respectively. These records provided a means for determining changes in ownership, number of parcels per family, and average parcel size over a period of approximately fifty-five years.

The second issue concerned the role of migration in mitigating conflict in the watershed. The approach was to compile a detailed history of one of the prominent families living within the river basin. The record compiled from oral histories extended back five generations and consisted of births, deaths, and migration and destination wherever possible. These data were used with other existing demographic data to evaluate the significance of this process over time.

The survey was also used, in part, to determine what policies were contributing to overgrazing of upland areas. Five policies, four extant and one newly proposed, were the focus of the study. The four extant policies were provincially set cattle export ceilings on Siquijor livestock to maintain local meat supplies, municipal ordinances regulating open grazing on privately held lands, and two national programs designed to increase herd size and improve pasture respectively in the provinces. Finally, the current administration's newly proposed comprehensive agrarian reform program (CARP) would transfer privately held abandoned lands to landless tenants.

In addition to the inclusion of specific questions in the questionnaire, reviews of relevant government documents at the provincial and community level were conducted. Finally, based on a preliminary analysis of these sources of data, interviews with local government officials and extension agents knowledgeable of the respective programs were conducted.

The approach to implementing the two research components was staggered, beginning with the biophysical data collection, which was scheduled to coincide with the expected onset of the rainy season. Once data collection was underway, this was followed by the survey questionnaire.

Chapter 5

RESULTS

The Biophysical Component

General Physiography

Many characteristics of the study site appear to be representative of the island's general physiography. The watershed, measuring an estimated 14.7 km^2 in area or some 5 percent of the total insular land surface, is elongated along the downstream axis and narrows at both termini. It is characterized by steep slopes and a minimum of low-lying flat land confined to small alluvial valleys and a narrow coastal plain (figure 2).[1] The basin's underlying geological formation begins with volcanics typical of the island's higher elevations before being replaced by a younger sedimentary sequence of calcareous shales and limestones typical of the undulating foothills, and finally terminates in a small coastal valley.[2]

Results from the upland farm erosion survey also appear to reflect the island's geological evolution and predominant pedogenic processes. Most of the the farms are located in an area where average slope is equal to or exceeds 40 percent and the surface is characterized by extensive impediments and severe to excessive erosion.[3] The weathering of the shales and

---

1 Some 68 percent of the river basin consists of slopes equal to or exceeding 18 percent with an estimated 64 percent of land lying between 100 and 500 m in elevation. Republic of the Philippines, Department of Agriculture and Food, Bureau of Soils and Water Management, *Slope and Elevation Maps, Province of Siquijor, 1:50,000* (Manila: Bureau of Soils and Water Management, 1985).

2 Republic of the Philippines, Ministry of Natural Resources, Bureau of Mines and Geo-Sciences, "Geological Verification of a Phosphate Quarry Area AQP-646(3) in Barangay Pisong-A, Municipality of Maria, Siquijor Island," by Alvin M. Matos (Cebu City: Bureau of Mines and Geo-Sciences, 1984): 4.

3 The occurrence of "slumps," small landslips in areas of steep slope, appear to be largely confined to the community of Calunasan. This phenomenon may be related to the increased presence of Faraon clay in the area of sampling, the stony nature of which contributes to faulting. The occurrence of these surface slips may represent a possible significant additional source of sediment when newly exposed soil is subject to erosional forces.

TABLE 4. Selected surface and soil characteristics from farm survey of six upland *barangays* in the Maria River basin, Province of Siquijor.

Surface characteristics (number of farms)

	Slope class (%)						Surface Impediments (%)[a]					Slumps[b]
Barangay	0-5	5-8	8-20	20-40	40-60	>60	0-20	21-40	41-60	>60	N	(no. obs.)
Basac	0	0	0	0	7	11		8	4	6	18	0
Bonga	0	0	0	0	0	1	0	1	0	0	1	3
Calunasan	0	0	0	2	10	5	0	4	9	4	17	24
Candigum	2	0	0	1	5	6	1	5	4	4	14	0
Catamboan	0	0	0	2	7	8	0	7	5	5	17	0
Pisong B	2	1	0	0	3	3	3	3	2	1	9	0

Soil characteristics (number of farms)

	Geology				Predominant soil type (clay)			Erosion class				
Barangay	Vo	Cs	Ls	Sh	Al	Lu	Fa	Sl	Mo	Sv	Ex	N
Basac	1	10	5	2	1	12	5	0	0	9	9	18
Bonga	1	0	0	0	1	0	0	0	0	0	1	1
Calunasan	5	2	10	0	5	2	10	0	0	11	6	17
Candigum	0	2	6	6	0	8	6	1	2	7	4	14
Catamboan	0	14	2	1	0	15	2	0	0	8	9	17
Pisong B	1	7	1	0	1	7	1	2	1	4	2	9

Note: Owing to logistical constraints associated with the soil survey, on-site sampling was limited to one in farm Bonga.

[a] Represents estimated percentage of land rendered nonarable due to physical outcroppings.

[b] "Slumps" describe small slope failures resulting in exposure of underlying subsoil or bedrock typically measuring .5-1.5 m in length by .5 m in width varying in depths.

Key:

Vo volcanics	Al alimodian clay	Sl slight
Cs calcareous shale	Lu Lugo clay	Mo moderate
Ls limestone	Fa Faraon clay	Sv severe
Sh shale		Ex excessive

limestone has resulted in the predominance of Lugo and Faraon clays in the watershed's upper reaches (table 4).[4]

Precipitation

Results from monitoring rainfall over the twelve-month study period demonstrate the orographic effect of the island's terrain as evident in the linear relationship between station means and altitude (table 5). Despite

---

[4] In addition a third soil type, alimodian clay, was identified. This is a low-fertility soil derived from noncalcareous shales.

TABLE 5. Average monthly rainfall in the Maria River basin, Province of Siquijor (in mm).

	Station (altitude above MSL)				Reference Station
	Cang-apa (380 m)	Calunasan (300 m)	Bonga (200 m)	Maria (10 m)	Mt. Bandilaan[a] (450 m)
July[b]	40.5	30.3	67.5	37.3	279.9
August	205.7	100.0	153.9	151.7	260.1
September	195.3	170.5	90.3	113.7	210.3
October	142.6	190.5	133.0	104.2	218.4
November	104.1	110.3	84.5	56.3	226.8
December	74.8	66.6	60.3	9.4	205.0
January	7.3	9.1	6.4	0.0	105.2
February	38.5	30.0	27.6	9.7	58.4
March	5.8	4.5	0.0	0.0	53.6
April	59.7	58.5	56.6	29.2	35.3
May	18.8	13.9	20.0	2.4	111.8
June	176.9	149.6	182.6	56.8	231.7
July[c]	39.1	61.5	38.8	12.4	-
Total	1109.1	995.3	921.5	583.1	1,811.5
Monthly station mean	92.4	82.9	76.8	48.6	151.0

Note: To date the only rainfall records that exist for Siquijor are those from one station located near the island's highest elevation (Mt. Bandilaan) for the period 1960-65 and a second station that began recording total rainfall in Cang-apa beginning in 1986. See University of San Carlos Water Resources Center, *Hydrology in Review Region VII* (Cebu City: Water Resources Center, 1984), p. 4-3.

[a] Figures based on an average five-year monitoring period. Height of gauge is approximate.
[b] Figures for the period July 15-31, 1987, only.
[c] Figures for the period July 1-14, 1988, only.

the paucity of data, the pattern generally conforms to the characteristics of Thornthwaite's subhumid class C, whereby rainfall diminishes during the months of February to April from maximum levels recorded from June to October.

A key characteristic affecting the values for the year of record was the presence of drought. While past records from the Mt. Bandilaan reference station are not strictly comparable owing to differences in altitude, they do appear to demonstrate the presence of severe drought, particularly in the months of July, December, and January of the study year. Further evidence of drought severity was reflected in the near-universal agreement among survey respondents describing the past year as one with a short rainy season characterized by infrequent and low-intensity rainfall. No statistically significant differences were observed in responses among upland and coastal resources users (table 6). The apparent recency and severity of the drought

TABLE 6. Comparative perceptions of climatic characteristics for 1987, Maria River basin, Province of Siquijor.

Questionnaire statement	Relative agreement of respondents							
	StD	D	SlD	N	SlA	A	StA	Total
The rainy season began early.								
Upland	15	54	31	0	1	1	0	102
Coastal	4	106	7	1	2	0	0	120
The rainy season began late.								
Upland	0	0	2	2	13	70	15	102
Coastal	0	0	1	2	2	107	8	120
The dry season began early.								
Upland	0	8	0	0	13	63	18	102
Coastal	0	0	0	2	3	107	8	120
The dry season began late.								
Upland	12	57	9	0	4	16	4	102
Coastal	3	105	9	2	1	0	0	120
The rains were more frequent.								
Upland	17	62	20	0	1	1	1	102
Coastal	3	103	9	2	2	1	0	120
The rains were less frequent.								
Upland	2	5	1	0	6	77	10	102
Coastal	0	1	0	2	5	102	10	120
The rains were more intense.								
Upland	10	59	30	0	1	2	0	102
Coastal	3	102	9	2	4	0	0	120
The rains were less intense.								
Upland	0	3	0	0	9	77	13	102
Coastal	0	0	0	2	6	102	10	120

Note: No statistically significant differences were observed in responses between upland and coastal managers for the eight statements (applying the Kruskal-Wallis test @ .1 significance level).

Key:
StD strongly disagree
D disagree
SlD slightly disagree
N neutral
SlA slightly agree
A agree
StA strongly agree

may in part explain some of the responses to questions concerning environmental perceptions.

Watershed Hydrology and Sediment Yield

The emphasis of the hydrological analysis was to ascertain the timing, significance, and nature of surface runoff reaching the coast. This in turn was useful in the characterization of the nearshore marine environment, particularly with regard to river-borne sediment and its influence on coastal fisheries. Daily water discharge data indicated that two distinct hydrological regimes associated with the seasonal monsoons occur in the lower reaches of the Maria watershed. During the southwest monsoon (June to October) when rainfall is abundant, discharge is relatively rapid. In contrast, during the drier months of the northeast monsoon (February to April), runoff is reduced. At the coast, reduction of flow is further affected by the formation of a partial sandbar attributed to the monsoon's westerly driven swell and serving to reduce the river's influence in the coastal waters. This had the added effect of masking runoff values determined from staff gauge observations (i.e., reduced flows created a pool of water behind the coastal sandbar resulting in artificially high river levels). As a result, two discrete sets of river gauge readings were recorded. Rather than attempt to calculate a constant to account for these two regimes, I prepared separate rating curves for each monsoonal period for use in subsequent calculations. Study year totals for water and suspended-sediment discharge and yield were calculated by prorating the respective values for the two regimes on the basis of total number of days recorded for each monsoon respectively. A flow duration curve was constructed for the two regimes (figure 5).[5] The upper ends of the two curves have been modified to account for extremely high but short duration flows.[6] Next, a sediment rating curve was constructed by plotting the results of the daily monitoring of suspended sediment and flow and fitted by eye (figure 6).[7] A sediment transport curve was then constructed based on the relationship between water discharge and suspended-sediment discharge at the time of sampling (figure 7). The points on the curve represent the suspended-sediment discharge at the time of sampling but are presented as if these "instantaneous" values remained constant throughout the course of the day.[8] Discharge for the monitoring year was calculated by dividing the total range of flow from figure 5 into smaller ranges and multiplying these subranges by the corresponding percentage of time that flow occurs for each monsoon (table 7). The sum of the products was multiplied by the number of days for each monsoon and then added together to represent a total study-year water discharge of 2,144 m^3/min.[9] By a similar approach, suspended-sediment load was determined

---

5 A flow duration curve is a cumulative frequency curve illustrating the percentage of time within the total period of record (in this case one year) that a specified daily discharge was equaled or exceeded.

6 See United States Geological Survey, *Reconnaissance Study of Sediment Transported by Stream Island of Oahu*, by B.L. Jones, R.H. Nakahara, and S.S.W. Chinn, Circular C33 (Honolulu: United States Geological Survey, 1971), pp. 21-23.

7 This represents an average relationship between water discharge and suspended-sediment discharge.

8 United States Geological Survey, *Reconnaissance Study*, p. 20.

9 This is equivalent to 35.73 m^3/sec., or 2,502.46 acre-ft./yr.

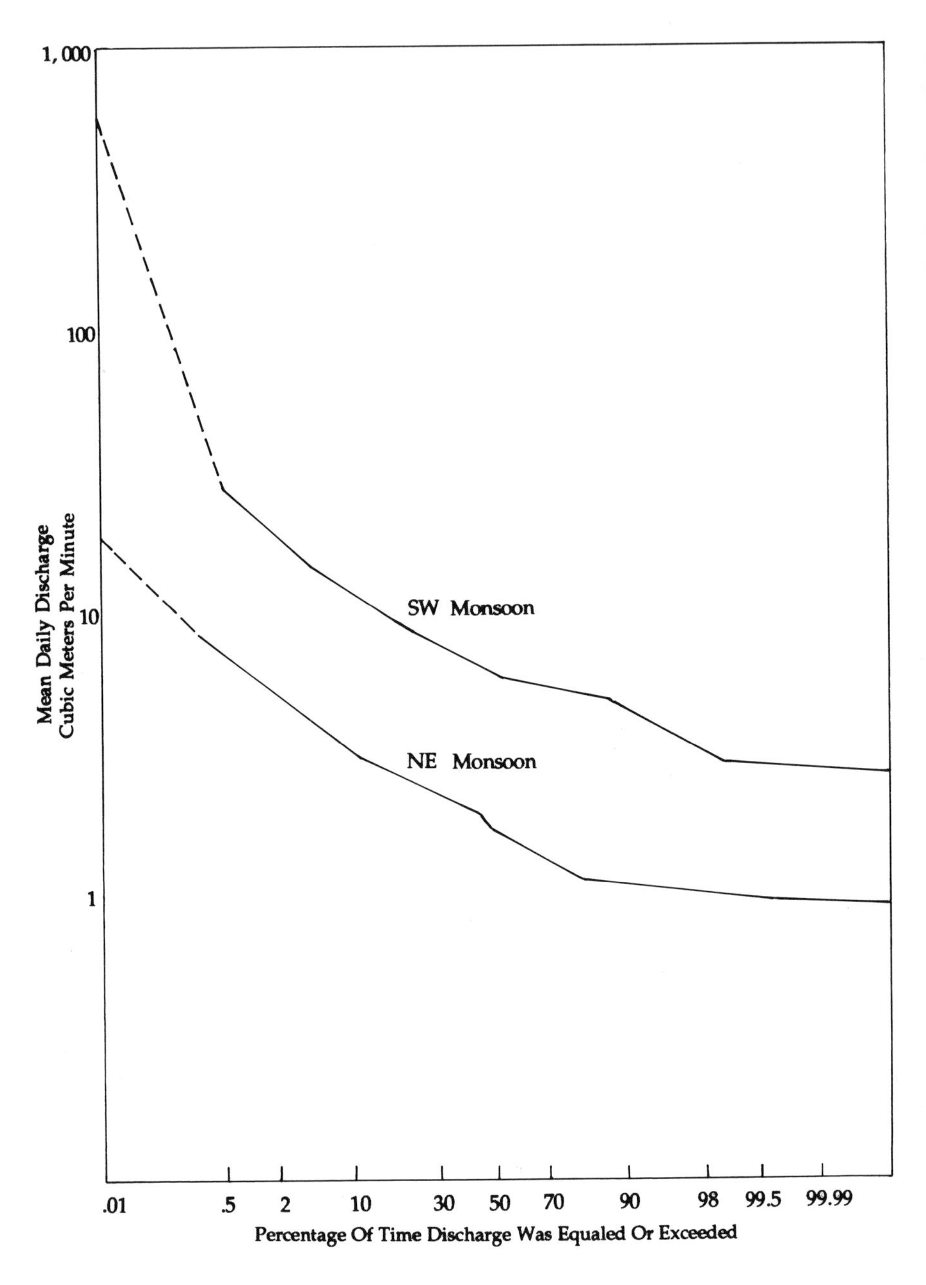

Fig. 5. Flow-duration curve for Maria watershed

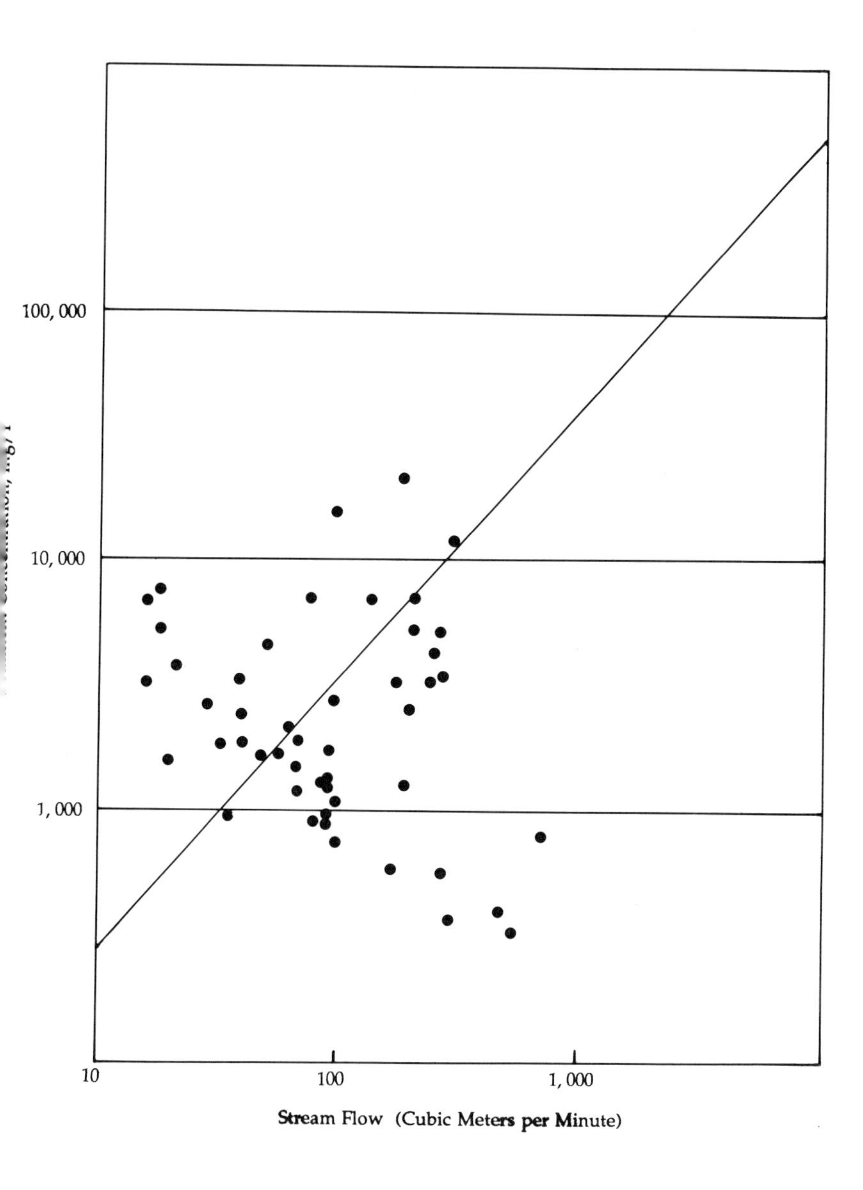

Fig. 6. Daily sediment rating curve for Maria watershed

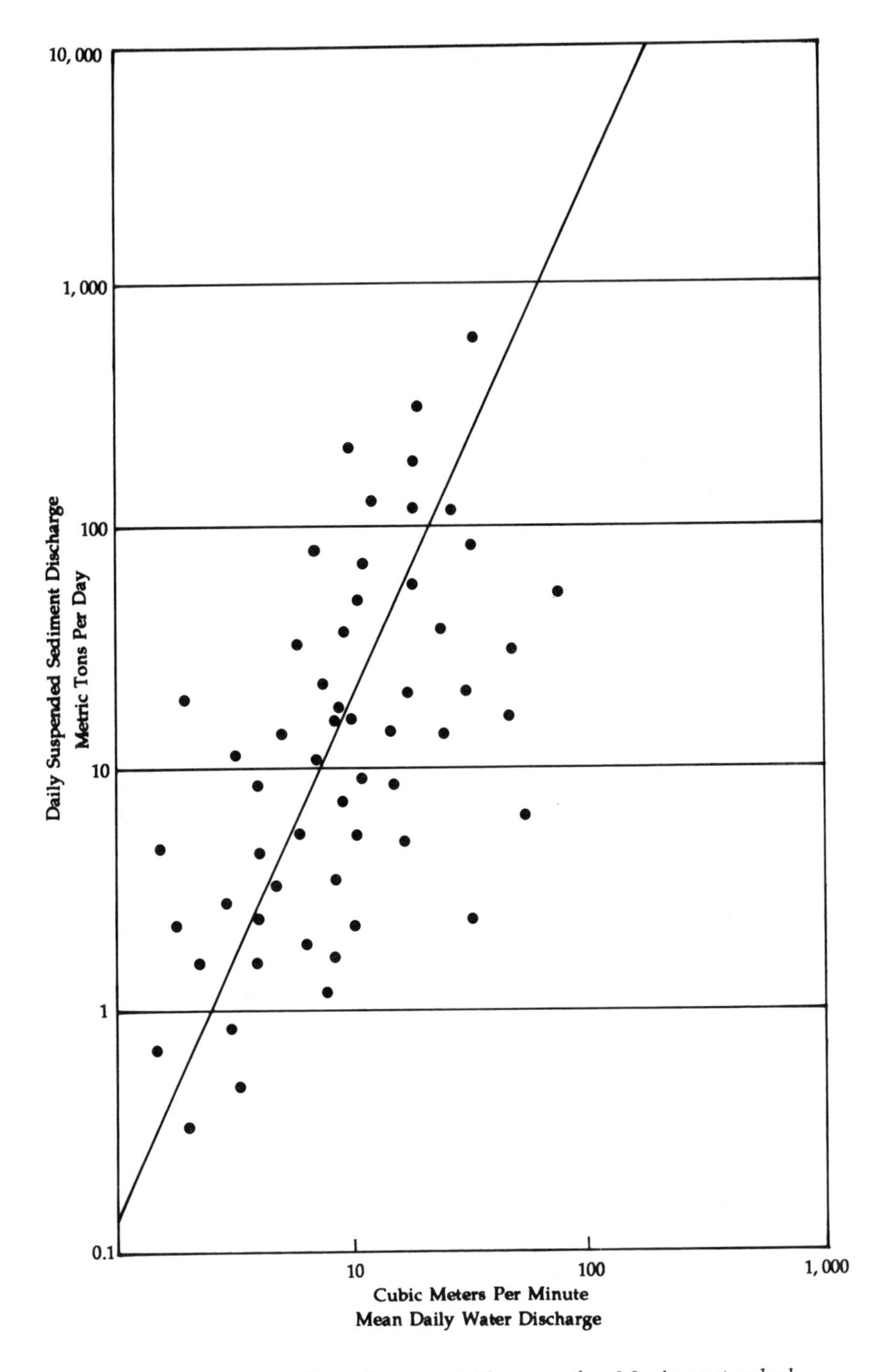

Fig. 7. Daily sediment yield curve for Maria watershed

for each water discharge range (from figure 7), multiplied by the corresponding percentage of time flow occurs, summed and prorated for the number of days for each monsoon. These calculations provided an estimate of annual suspended-sediment load and yield of 9,391 mtons/yr. and 639 mtons/yr./ km^2 respectively.[10] As a first estimate this figure is not unreasonable in light of the aforementioned conditions. However, it must be noted that this estimate was based on results obtained in an extremely dry year and does not account for bedload. There exist only a few rivers that have been gauged in Region 7 and none in Siquijor. As a result no sediment yield data exists for any of the island's watersheds. However, data exist from both the Buhisan and Mananga river basins on the island of Cebu. In the former, yield was estimated to be approximately twice that of Maria for the period 1951-77 (1,500 mton/km^2/yr.), the latter some eleven times the Maria figure (7,100 mton/km^2/yr). Obviously the variables involved are so numerous and complex that comparisons are of little value.[11]

A total of 106 rainfall events were recorded in the upland portion of the watershed (Cang-apa rainfall gauge). Of these, twenty-four were designated as significant, meaning that runoff in two nearby monitoring weirs was sufficient to crest the weir notch required for recording water flow readings. The average duration of the twenty-four events was 103 minutes (although the highest intensity rainfall was confined to periods approximating a half hour). Of the twenty-four events, thirteen resulted in discernible downstream flood peaks at the Maria River mouth monitoring station. The profiles of these peaks were classified into three types: (1) small, short peaks approximating an hour in duration occurring some thirty minutes after onset of rainfall (figure 8, type a); (2) a much sharper spike in the flood peaks lasting about an hour in duration occurring some sixty minutes after onset (type b); and (3) peaks characterized by both longer duration and sooner onset typical of larger scale events (type c).

Coastal Water Quality

As measures of water quality, both salinity and visibility are easily measured and often highly correlative with terrestrial runoff. They are also important environmental parameters to which a number of highly productive marine ecosystems are particularly sensitive, including coral reefs and

---

[10] An area of 14.7 km^2 was used for the calculation of yield. This is equivalent to 1,820 tons/yr./mi^2.

[11] C.J. Dystra, "Engineering Geological Feasibility of Rehabilitating Buhisan Reservoir," report prepared for the University of San Carlos Water Resources Center and the Delft University of Technology Cooperative Project, Cebu City, 1977 (typewritten), pp. 1-109; and Virgilio A. Sahagun, *Sediment Transport Study: Mananga River* (Cebu City: University of San Carlos Water Resources Center, 1985), unpaged.

TABLE 7. Duration table of mean daily discharge at Maria River mouth and corresponding

Limits of range (%)	Interval (%)	Midordinate (%)	Mean flow/discharge in range: Water m^3/min/d	Mean flow/discharge in range: Sediment mton/d	Total daily flow/discharge in range: Water m^3/min/d	Total daily flow/discharge in range: Sediment mton/d
Southwest monsoon period						
.00-.02	.02	.01	530	80,200	.11	16.04
.02-.05	.03	.035	290	20,290	.09	6.09
.05-.1	.05	.075	140	5,000	.07	12.50
.1-.4	.3	.25	50	550	.15	1.65
.4-.6	.2	.5	29.5	178	.06	.36
.6-1.0	.4	.8	28	162	.11	.65
1.0-1.4	.4	1.2	22	96	.09	.38
1.4-1.8	.4	1.6	21	89	.08	.36
1.8-2.2	.4	2.0	19.5	75	.08	.30
2.2-3.4	1.2	2.8	18	63	.22	.76
3.4-4.6	1.2	4.0	16	48	.19	.58
4.6-7.4	2.8	6.0	14	39	.39	1.09
7.4-11	3.6	9.2	13	31	.47	1.12
11-15	4	13	11	22	.44	.88
15-25	10	20	9.3	15	.93	1.50
25-45	20	35	7.2	8.7	1.44	1.74
45-65	20	55	6.0	6.0	1.2	1.2
65-85	20	75	5.3	4.6	1.06	.92
85-95	10	90	4.6	3.4	.46	.34
95-100	5	97.5	3.6	2.0	.18	.1
Totals					7.82	38.56

Total water discharge x 241 days of southwest monsoon = 1,885 m^3/min

Total sediment discharge x 241 days of southwest monsoon = 9,293 mtons

1987-88 total discharges

Water = 1,885 + 259 = 2,144 m^3/min. (= 2,503 acre-ft./yr.)

Sediment = 9,293 + 98 = 9,391 mtons

Suspended-sediment yield = 9,391 mtons/14.7 km^2 = 638.8 mtons/km^2/yr. (= 1,820 tons/mi.^{2}/yr.)

marine grassbeds. Many coral species are stenohaline and together with marine grasses are also sensitive to turbid waters. Turbidity, stemming from high sediment concentration, can affect corals and marine grasses negatively. This occurs principally through the reduction in light penetration, which interferes with photosynthesis, resulting in a lowering of growth rates. Moreover, several literature reviews indicate that sedimentation can suspended-sediment discharge , July 1987 to July 1988, Province of Siquijor.

			Mean flow/discharge in range		Total daily flow/discharge in range	
Limits of range (%)	Interval (%)	Midordinate (%)	Water m^3/min/d	Sediment mton/d	Water m^3/min/d	Sediment mton/d
Northeast monsoon period						
.00-.02	.02	.01	19	70	.01	.01
.02-.05	.03	.035	17	58	.01	.02
.05-.1	.05	.075	12	26.4	.01	.01
.1-.4	.3	.25	8	11	.02	.03
.4-.6	.2	.5	7.2	8.7	.01	.02
.6-1.0	.4	.8	6.6	7.5	.03	.03
1.0-1.4	.4	1.2	6	6	.02	.02
1.4-1.8	.4	1.6	5.3	4.5	.02	.02
1.8-2.2	.4	2.0	5.1	4.4	.02	.02
2.2-3.4	1.2	2.8	4.8	3.7	.06	.04
3.4-4.6	1.2	4.0	4.1	2.6	.05	.03
4.6-7.4	2.8	6.0	3.8	2.5	.11	.07
7.4-11	3.6	9.2	3.1	1.4	.11	.05
11-15	4	13	2.9	1.3	.12	.05
15-25	10	20	2.7	1.1	.27	.11
25-45	20	35	2.1	.55	.42	.11
45-65	20	55	1.8	.36	.36	.07
65-85	20	75	1.3	.23	.26	.05
85-95	10	90	1.2	.19	.12	.02
95-100	5	97.5	1.1	.16	.06	.01
Totals					2.09	.79

Total water discharge for northeast monsoon x 124 days = 259 m^3/min

Total sediment discharge for northeast monsoon x 124 days = 98 mtons

affect coral reefs adversely, either by reducing growth rates or in certain cases causing outright death, typically through inhibition of gas exchange feeding ability.[12] In the present study, visibility (the measure of light penetration) was used and interpreted as a surrogate measure of water turbidity.

From an analysis of the salinity and visibility results, one can discern several patterns. For the study's purposes, differences in monthly salinity

---

12 However, effects vary among coral species owing to the differing influence of particle size on the physical attributes of the coral polyps. See R. Endean, "Pollution of Coral Reefs," in *5th FAO/SIDA Workshop on Aquatic Pollution in Relation to the Protection of Living Resources* (Rome: United Nations Food and Agriculture Organization, 1978), pp. 343-363; and E.J. Ferguson Wood and R.E. Johannes, eds., *Tropical Marine Pollution*, Elsevier Oceanography Series 12 (Amsterdam: Elsevier Scientific Publishing Co., 1975), pp. 15-18.

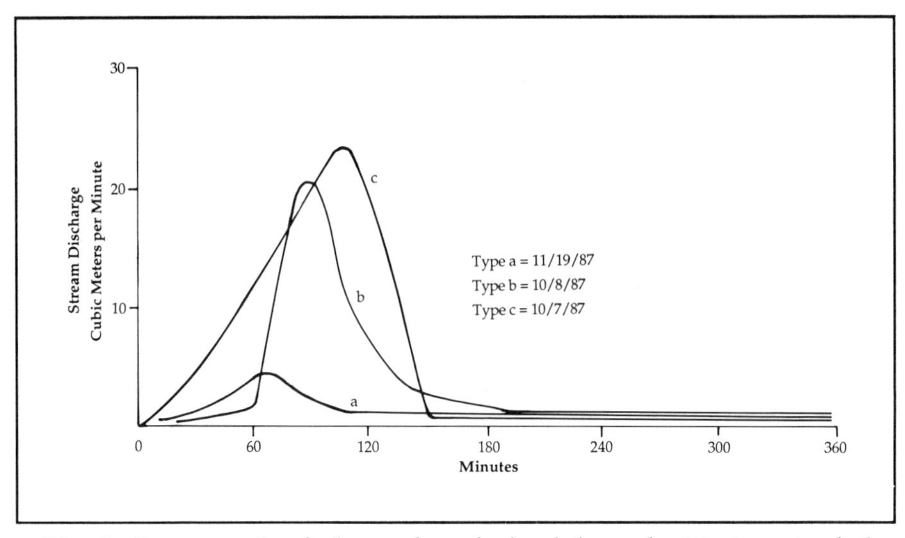

Fig. 8. Representative hydrographs calculated from the Maria watershed

values among the five monitoring stations were insignificant (table 8). However taking the bay as a whole, a seasonal pattern in salinities appears. For the greater part of the year, values are lower than open ocean salinities (thirty-five parts per thousand). This is attributable to the combined effects of the island's surface and subsurface drainage. During the early onset of the rainy season, lowered salinities are most likely a result of surface runoff. As rainfall diminishes toward the end of the season, groundwater discharge continues into the bay, owing to the time lag associated with percolation and groundwater recharge, a process one expects to be significant in the limestone-dominated bedrock. Groundwater influence continues until it too diminishes in the absence of replenishment, resulting in gradual increases in salinity values.

TABLE 8. Average monthly surface salinity (in parts per thousand), Maria Bay, Province of Siquijor, October 1987 to July 1988.

Station	1987 Oct.	Nov.	Dec.	1988 Jan.	Feb.	Mar.[a]	Apr.[a]	May	June	July
3	30.7	30.4	29.9	29.9	29.1	-	-	33.4	32.9	34.9
7	30.8	30.5	29.8	29.9	29.7	-	-	33.6	32.5	34.9
4	30.6	30.2	29.9	29.9	29.3	-	-	33.5	33.0	35.0
5	30.5	30.3	29.9	30.1	29.9	-	-	33.5	32.8	35.2
6	30.7	30.3	29.8	29.7	29.2	-	-	33.5	32.8	35.1
Average	30.7	30.3	29.9	29.9	29.4	-	-	33.5	32.7	35.0

[a] Values not obtained.

In contrast, a clear spatial pattern can be discerned in the visibility data (table 9). Visibility is lower in the four southernmost stations (7, 4, 5 and 6), an area of the bay with the greatest river density (figure 3). This conclusion is further supported by comparing the lowest and highest visibility values from stations 4 and 3 respectively, the former directly in front of the mouth of the Maria River, the latter located in an area of the bay characterized by an absence of well-defined drainage.

TABLE 9. Average monthly water visibility (in meters from surface), Maria Bay, Province of Siquijor, October 1987 to July 1988.

Station	1987 Oct.	Nov.	Dec.	1988 Jan.	Feb.	Mar.	Apr.	May	June	July[a]	Avg.
3	9.1	8.4	6.6	10.3	10.9	10.1	11.0	11.4	11.0	11.0	10.0
7	8.1	7.6	5.2	8.4	8.6	8.1	8.5	8.6	8.1	8.1	7.9
4	6.9	6.2	4.6	6.4	7.0	6.3	6.5	6.7	6.5	6.4	6.4
5	9.0	7.0	4.8	7.5	7.2	7.6	8.0	7.8	7.7	8.0	7.5
6	9.2	8.3	5.3	6.4	6.7	6.4	6.8	7.0	7.2	6.9	7.2
Avg.	8.5	7.5	5.3	7.8	8.1	7.7	8.2	8.3	8.2	8.1	7.8

[a] Values based on July 1-14 only.

Not surprisingly this pattern is reflected in the sediment trap data. Observed values are high from stations in most direct proximity to the rivers' mouths and are at their lowest in the southern and northernmost stations, particularly the latter (table 10). Moreover, values increase during the months of November through March, in some cases by an order of magnitude.

These patterns in the visibility and sedimentation data are indicative of the two principal processes contributing to bay sediment. In the first process, bay turbidity appears primarily to be a result of the transport of terrigenous material through river runoff, no doubt at its highest levels during the rainy season months. However, with the onset of the northeast monsoon, a period characterized by low rainfall but strong northeasterly winds, reduced visibility is thought to be due to the resuspension of bottom sediments attributed to the increased wave energy pounding Maria's exposed coast.

Currents

Understanding the bay's current regime was important in assessing the areal influence of river-borne sediment on benthic communities and in turn, possible locational preferences of fishermen. However, because of the previously described problems in the conduct of current studies during the northeast monsoon, an insufficient number of observations were com-

TABLE 10. Average daily rate of sedimentation (in centigrams/cm^2), Maria Bay, Province of Siquijor, July 1987 to July 1988.

Station	1987 July[a]	Aug.[a]	Sept.	Oct.	Nov.	Dec.[b]	1988 Jan.	Feb.	Mar.	Apr.	May	June	July	Avg.
3	.78	.97	1.12	1.12	1.93	3.91	1.26	.07	.75	.77	.78	1.84	-[c]	1.28
7	1.54	1.34	1.12	1.12	4.24	12.18	12.36	10.17	13.56	11.60	2.35	1.62	1.64	6.24
4	5.01	8.59	11.56	10.07	10.40	85.68	20.22	242.70	113.56	53.66	7.51	3.41	4.19	44.35
5	18.88	15.50	11.93	27.21	34.67	68.94	90.08	179.24	203.76	123.33	10.17	9.90	14.51	62.16
6	6.55	5.36	4.10	11.18	9.25	14.92	12.36	-[c]	32.37	50.68	-[c]	0.00	0.00	16.31
Avg.	6.55	6.35	5.97	10.14	12.10	37.13	27.26	108.04	72.80	48.00	5.20	4.20	6.80	

a Monthly figures for July and August were extrapolated from the monitoring period covering July 15 through August 31.
b December and January figures for station 6 were extrapolated from the monitoring period covering December 1 through February 3.
c Sample not recovered.

pleted to develop a sound understanding of the bay's current pattern regime. Despite this deficiency, some insights can be gleaned from the eight discrete studies monitoring six combinations of tidal phases and conditions that were completed (table 11). Further, studies 1, 1a, and 2 were conducted under conditions approximating the northeast monsoon (owing to the recent passage of a depression characterized by strong winds and swell running from the northeast) and were treated as such. Based on these results and the previously stated caveat, the following processes appear to be at work. Regardless of tidal conditions, there appears to be a discernible northerly flow on the tide's ebb as indicated in tidal studies 1 and 3 (figure 9). While study 1a was also conducted under these conditions and produced results indicating a southerly flow, its initiation was sufficiently close to the predicted time of the flood phase that this could be explained by premature onset.[13] In contrast, the flood condition appears to result in a southerly and westerly flow with the exception of study 6. One possible explanation for the latter's long excursion to the northern extremity of the bay was the occurrence of strong southerly winds and swell during the study, conditions unique to this study.

Velocity, graphically reflected in absolute straight-line distance traveled by the drogue, was generally weak, with excursions confined to an arc of 350 m in front of the Maria River mouth in five of the eight studies. The exceptions were the two long excursions to the north of the bay, reaching up to 3 km, and one southerly excursion extending some 780 m. The studies indicate that under long semidiurnal and most probably diurnal ebb events (which were not studied) or under conditions of strong southwesterly wind and swell, currents are capable of traveling (and transporting materials below threshold densities) the length of the bay as measured from the river mouth to the bay's northernmost extremes. In contrast, long period flood phases, though resulting in a southerly flow, do not generate currents of equivalent strength. In the absence of these conditions, currents appear relatively weak, alternating between northerly and southerly directions during the ebb and flood phases respectively. Northeast monsoon conditions, as represented by the conditions in studies 1 and 1a, do not appear to affect these patterns significantly.

Substrate and Biotic Zonation

Results from the sampling of bottom substrate predictably reflected the interaction of the previously described processes. The northern and southern portions of the bay are characterized by a carbonate sand bottom

---

13 Siquijor's tidal conditions and times are only predictions based on a reference station located in Cebu.

TABLE 11. Currents, Maria Bay, Province of Siquijor, 1987.

	Study[a]							
	1	1a	2	3	4	5	5a	6
Date	10/7	10/7	10/7	10/8	10/8	10/15	10/15	11/26
Monsoon[b]	SW[c]	SW[c]	SW[c]	SW	SW	SW	SW	SW
Tidal condition[d]	SD	SD	SD	SD	SD	D	D	SD
Tidal phase	ebb	ebb	flood	ebb	flood	flood	flood	flood
Study time (min.)	270	—	199	411	340	908	—	440
Distance traveled (m)	400	290	320	3,290	1,530	380	4,190	3,280
Absolute distance[e] (m)	220	300	170	2,390	250	340	780	2,920
Direction[f]	N	SW	SW/W	NNE	SW/NW/SE/W/N	SW	SW/NE	NNE
Rate (m/min.)[g]	1.7	4.8	1.6	7.8	4.5	2.2	5.8	7.5
Wind direction	NE	NNE	NE	NNW	NE/WNW	NE/SW/NW	SSE	SSW/SSE
Wind speed (knots)	7-9	5-7	6-8	2-3	2-4	4-6, 7-9	4-6, 2-4	10-12, 4-8
Swell direction	ENE	ENE	ENE	ESE	ENE	ENE/SW	NNE/NW/SW	SE/SSE
Swell speed (c/sec.)[h]	3-4	5-7	3-4	3-4	3-4	2/2	2/2	3-4 / 3-4
Swell height (m)	.5-1	.5-1	.5-1	.5-1	.25-.5	.1 / .1	.1 / .1	.3-.5 / .3-.5

a Studies designated with a small letter identify subsequent monitoring under the same conditions following the termination of an event, usually owing to the drogue running aground.

b Designates conditions characteristic of the northeast or southwest monsoon.

c Despite the occurrence of the southwest monsoon, wind and swell conditions approximated those more characteristic of the northeast monsoon owing to the passage of a storm front.

d Refers to diurnal (D) or semidiurnal (SD) conditions.

e Represents distance in a straight line between the station and the point of study termination.

f Points of the compass were divided into twelfths (e.g., 0-30° is NNE; 30-60° is NE).

g Indicates average velocity, calculated from total distance divided by total time.

h Calculated by counting number of seconds between swells.

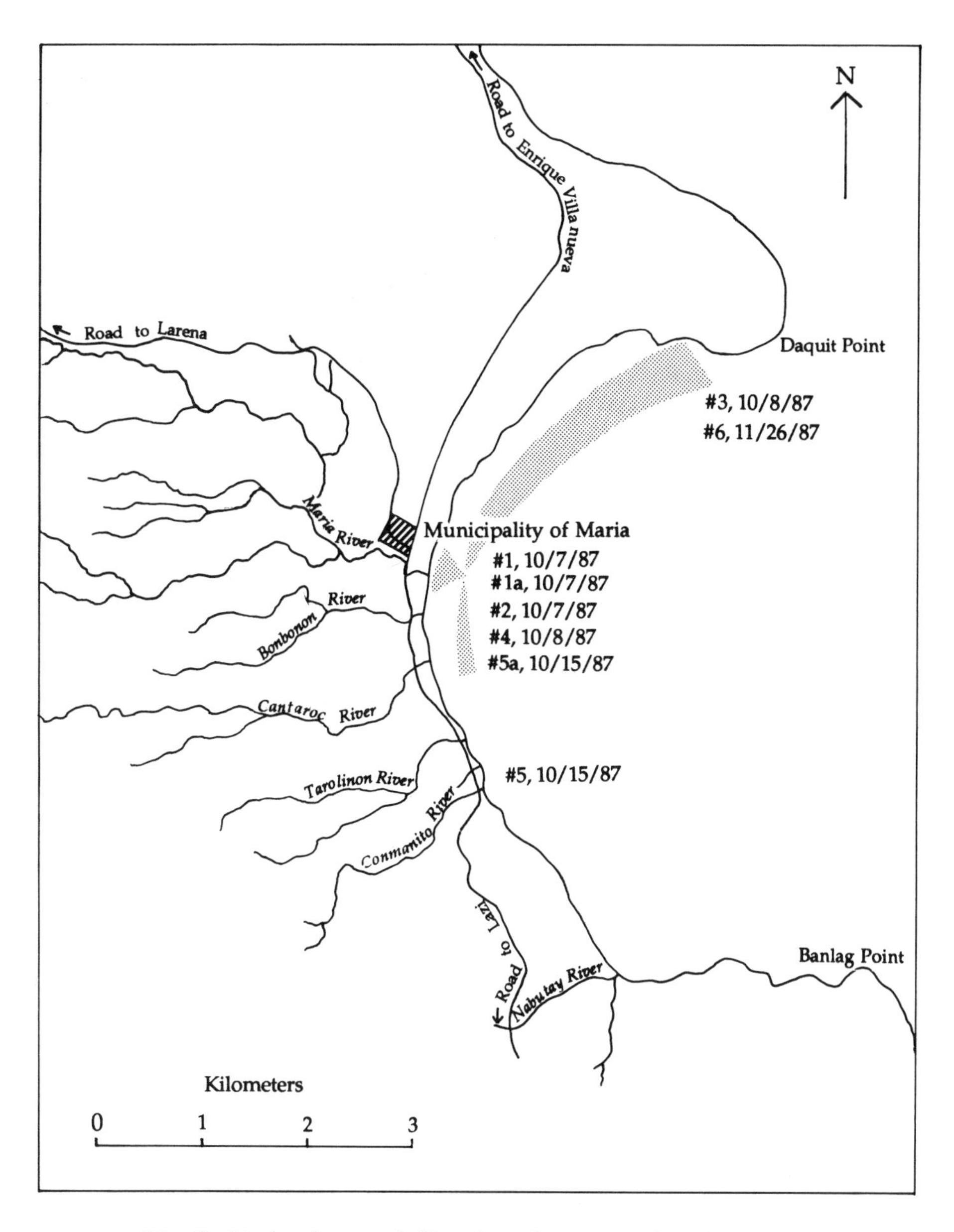

Fig. 9. Study, date, and direction of currents, Maria Bay

and interspersed patches of hard bottom dominated by coral reef (figure 10). In contrast, within proximity to the Maria River mouth and the other rivers to the south, the bottom offshore is entirely "soft," with sand turning into mud with distance offshore where river flow velocities are reduced and finer grain sediments settle out.

Biotic zonation patterns follow to a large degree substrate zonation. Sea grass is largely confined to the southern (and less turbid) extremity of the soft substrate zone and a few isolated patches (figure 11). Stressed coral reef, defined here as areas of hard bottom where the ratio of live to dead reef is less than one, occurs primarily in the south central portion of the bay near the rivers on the perimeter of the soft substrate zone (figure 11). Not surprisingly, healthy reef (ratios > 1) is found in the northern and southern section of the bay at greater distances from the influence of the rivers.

Fish Abundance and Diversity

A total of ninety-one species of fish representing twenty-five families were identified in the fish censuses (Appendix 3). Species abundance and composition were dominated by five families (figure 12). Estimates of the mean number of species and of the mean number of individuals per 500 m^2 were 20.5 and 773 respectively. These figures are indicative of a fish population low in both numbers and diversity. Data from a study assessing the effects of protective management of corals on fish populations from three sites in the Central Visayas indicated that total fish abundance exceeded the study's estimates by two orders of magnitude and recorded higher species diversity among all targeted fish families.[14]

The Socioeconomic Component

General Community Characteristics

The Upland Farming Community. Upland farming in the Maria is dominated by the planting of corn. Corn harvests occur twice or even three times a year depending on the amount of rain.[15] In addition to traditional crops, the grazing of livestock dominated by cattle, carabao, and goats is omnipresent on the hillsides, contributing to the severe erosion discussed above. The results from a number of survey questions were compiled to develop a profile of an upland farmer typical of the sample population (table 12).

---

14 See Garry Russ, "Effects of Protective Management on Coral Reef Fishes in the Central Philippines," in *Proceedings of the Fifth International Coral Reef Congress* 4 (1985): 219-224.

15 Triple cropping of corn has been reported as early as the 1930s and was known to occur on most farms even then. See Rosell, "Siquijor," p. 436.

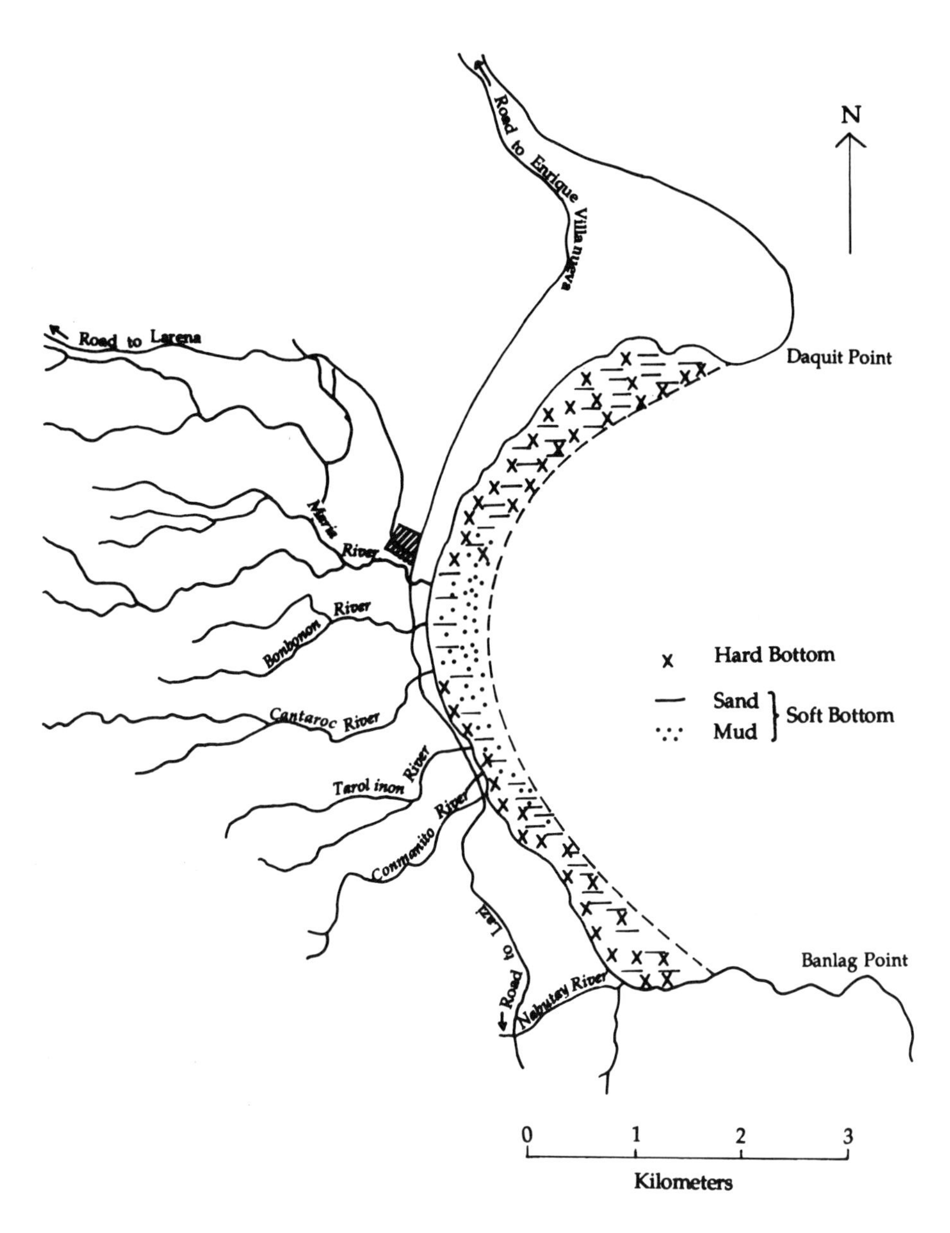

Fig. 10. Predominate substrate type, Maria Bay

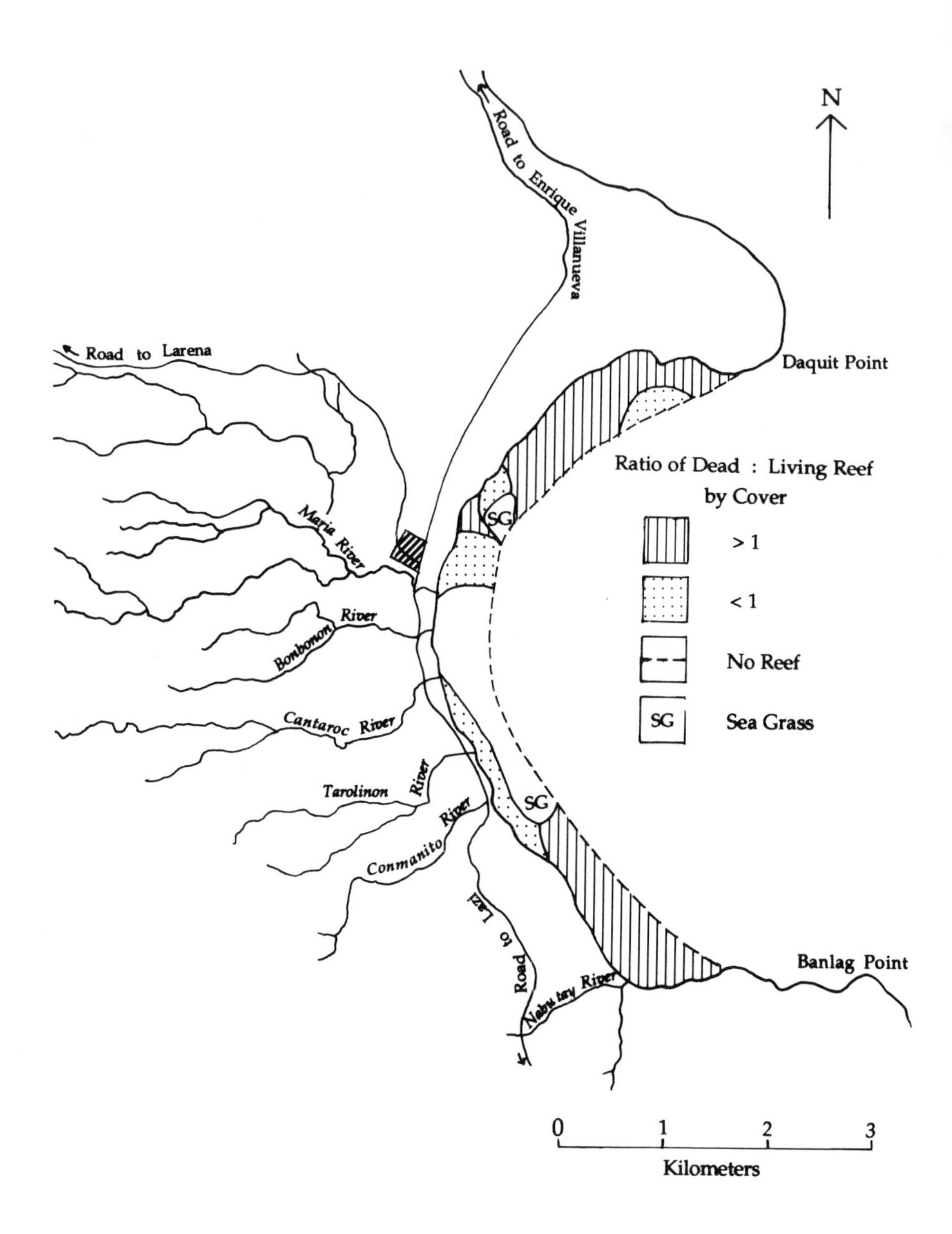

Fig. 11. Distribution and status of coral reef, Maria Bay

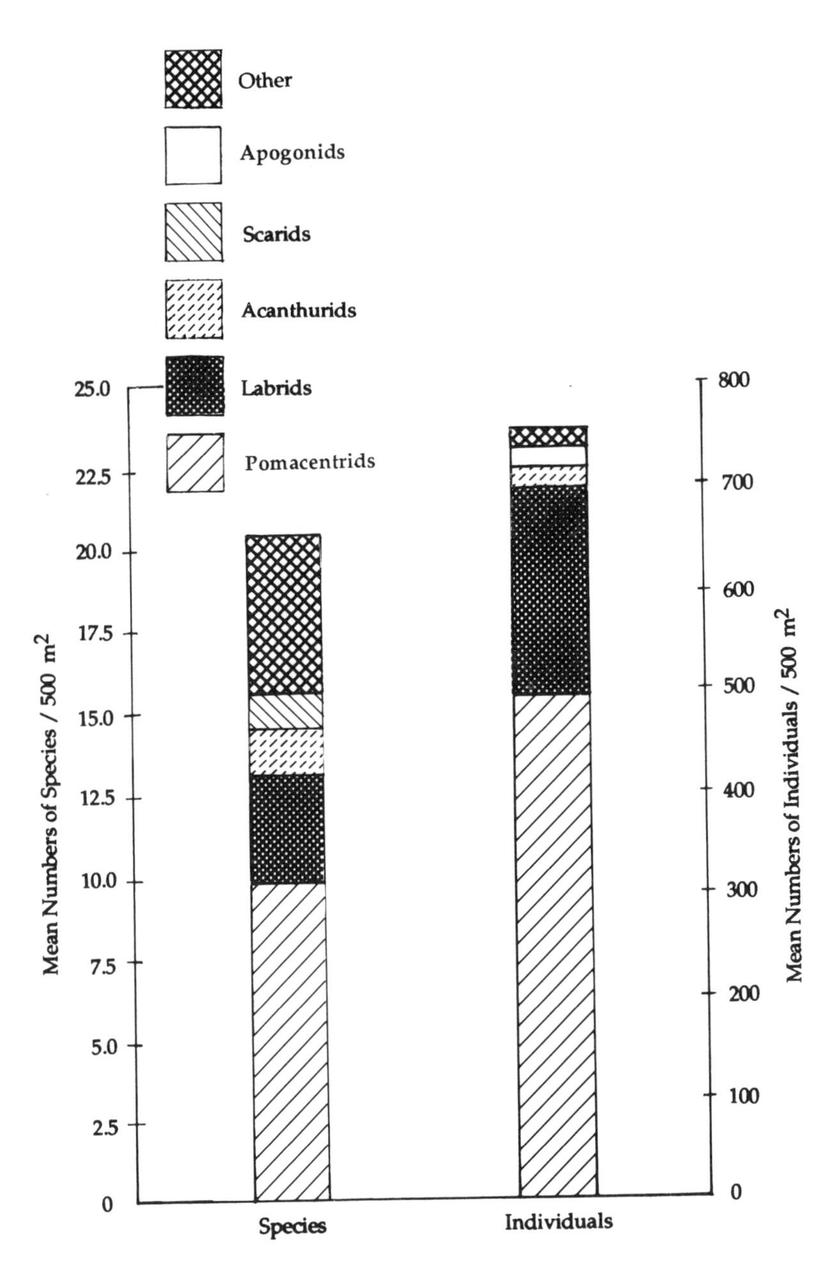

Fig. 12. Mean number of fish species and individuals of coral reef fishes per 500 m^2 area, Maria Bay

TABLE 12. Selected characteristics of the upland farming community in the Maria River basin.

Personal	
Mean age of farmer	48
Mean number of years in farming	19
Mean number of children	3.9
Mean number of siblings in communally owned lands	5.4
Farm	
Mean farm size (102 farms)	2.92 ha
Mean pasture size (86 pastures)	1.23 ha
Pasture:total farm area	36%
Livestock	
Mean head of cattle/farm[a]	1.89
Mean head of carabao/farm[a]	.29
Mean head of goats/farm[a]	.67
Mean number of cattle/farm ha	.66
Mean number of cattle/pasture ha	1.82
Mean number of cattle and carabao/farm ha	.75
Mean number of cattle and carabao/pasture ha	2.1

Note: Figures pertain to all individually operated farms (owned, partially owned, or tenanted).

[a] Includes all farms surveyed.

A set of questions attempted to establish trends in farm:pasture:cattle ratios over the past fifteen years. Farm size appears to have remained largely static; where changes have occurred they have been expansive typically for purposes of increasing production or preserving land for future generations (table 13). In contrast the predominant view with respect to pasturelands was fairly evenly divided between no change and diminishment in area grazed, the latter attributed chiefly to drought, increasingly unproductive lands, and competing land uses. Where increases were noted, they were a byproduct of other farmers abandoning their lands. Even more farmers stated that their herd size was decreasing over past years, a result of sales to meet a range of expenses and/or herd attrition (old age, sickness, accidental death). Probing in greater depth, I found that the typical upland cattle owner has sold one or more head at least once in his life, for purposes of meeting expenses that may range from paying for education to sponsoring a local fiesta (table 14).

The survey attempted to quantify recent trends in grazing density on owner-operator farms (table 15). The results indicate that the mean size of farms declined from 3.3 to 2.5 ha, while pasture area remained approximately the same, and cattle herd size fluctuated between 2.2 and 3 head per farm. Based on these figures no recent trends could be clearly identified in

TABLE 13. Temporal changes in selected farm characteristics of the Maria River basin's upland farming community.

	Farmland	Pastureland	Herds
Farmers increasing total size of holdings	11	13	9
Farmers decreasing total size of holdings	0	39	55
Farmers whose holdings remained the same size	91	35	20
N/A	0	15	18
Totals	102	102	102

Rationales for increase in farm size		Rationales for change in pasture size	
To increase production	4	Reductions	
For future generations	3	Unproductive lands	10
To keep in family	1	Toxic grasses	3
To build residence	2	Effects of drought	18
To assist in-laws	1	Competing land uses	8
Other	1	Increases	
Total	12	Adjacent vacant lands	10
		Other	3
		Total	52

Rationales for change in herd size			
Reductions		Increases	
To meet expenses	28	Power inputs	2
Herd attrition	5	Investment for future	1
Danger of occupation	5	Replace old or sick stock	1
Tenant obligations	2	Other	5
Personal consumption	2		
Expenses and attrition	14	Total	65

the aforementioned characteristics or the cattle:pasture ratios over the fifteen-year time period. This contrasts with the trends documented in parcel fragmentation data for Cang-apa over a much longer time period (see below).

Despite Siquijor's high percentage of landownership, the percentage of upland farmers owning their land (either a portion or in its entirety) is only just above parity in comparison to the number of tenants (table 16). Tenure arrangements between tenant and owner vary, but a 66:33 division predominates. A clear bimodal split occurs between those farmers who own their pasture outright and those asserting communal grazing rights (defined by the right to graze cattle freely on adjacent lands). Where cattle are not owned in their entirety, offspring are divided on a 50:50 basis between owner and tenant. The roles of the Department of Agriculture and CVRP appear to be minimal with regard to cattle distribution.

TABLE 14. Commercial importance of cattle in the Maria River basin upland farming community.

Age of cattle at sale		Frequency of sales[a]		Purpose of sale[b]	
<1 year	-	Never	12	To meet expenses[c]	48
1-2 years	7	Once	21	Cattle incapacitated	1
2-3 years	14	Twice	12	Danger to personal safety	6
3-4 years	24	3 times	9	Old age of cattle	7
4-5 years	16	> 3 times	15	N/A	45
5-6 years	15	N/A	33		
6-7 years	16				
>7 years	12				
N/A	45				
Mean age	4.1 years				

[a] Over the course of a farmer's career.

[b] Some respondents named more than one purpose.

[c] These include a wide range of needs, such as education, household goods, health costs, farm goods, the purchase of a carabao, and the costs of local fiestas.

Coastal Livelihood Groups. Land and water use on the coast is dominated by irrigated paddy farms, coconut production, and small-scale fishing as sources of livelihood. Paddy farms are mostly under 1 ha in size and limited to one or two croppings per year except where irrigation and short-term varieties can support a third crop. In the Maria watershed these irrigation schemes are small, communally owned and maintained systems serving to transport spring water to the system of paddy farms.

Coconut production coexists with most other land uses and is family owned and operated.

Finally, within the municipality of Maria there are an estimated 770 households dependent to varying degrees on fishing (of the 770 only 120 considered themselves to be full-time fishermen).[16] The fishery is typical of Siquijor, dominated by subsistence fishermen.

For purposes of characterizing the coastal community's principal sources of income, the three livelihood groups were further divided into individuals with no means of support other than from one or more of the three livelihoods, and those who received such additional income (table 17). Most individuals have at least one outside source of income, particularly among those persons who depend primarily on only one of the three main livelihoods. However, multilivelihood dependency is highly prevalent and appears to blur distinctions between the groups.

---

[16] Republic of the Philippines, National Economic and Development Authority, National Census and Statistics Office, *1980 Census of Fisheries: Region VII-Central Visayas*, vol. 1, *Final Report* (Manila: National Census and Statistics Office, 1980), p. 13.

TABLE 15. Changes in farm, pasture, and herd size over time in the Maria River basin's upland farming community.

	Area in ha									
	0-1/8	1/8-1/4	1/4-1/2	1/2-1	1-2	2-3	3-5	>5	$\overline{X}$	Total
Farms										
Present	2	10	7	8	13	7	3	9	2.54	59
5 yrs ago	1	10	7	12	5	5	3	7	3.31	50
10 yrs ago	1	7	8	5	8	4	2	7	3.07	42
15 yrs ago	1	7	5	5	7	2	2	8	3.28	37
Pastures										
Present	17	4	8	7	2	2	2	4	1.44	44
5 yrs ago	15	4	5	5	3	2	3	3	1.27	40
10 yrs ago	12	3	3	2	1	-	1	2	1.31	24
15 yrs ago	10	4	2	1	1	-	1	2	1.45	21

	Number of cattle							
Cattle herds	1	2	3	4	5	>5	$\overline{X}$	Total
Present	30	25	15	5	5	3	2.24	83
5 yrs ago	11	4	16	15	11	10	3.01	67
10 yrs ago	5	2	6	2	3	16	2.69	34
15 yrs ago	2	2	-	3	2	11	2.40	20

	Cattle per ha	
Cattle:land relationships	Farms	Pastures
Present	.88	1.56
5 yrs ago	.91	2.37
10 yrs ago	.88	2.05
15 yrs ago	.73	1.66

Note: All values in table pertain only to owner-respondents of land, pasture, and cattle.

Community Perceptions

Trends in Well-Being. As a first step to defining perceptions in both upland and coastal communities, questions were asked about the respondents' general economic situation and more specific status of livelihood and productivity. Responses to a suggestion that their situation was improving economically indicated that upland farmers were decidedly pessimistic; approximately two-thirds of the respondents indicated slight to general disagreement with the statement. In contrast, paddy farmers and coconut producers were neutral or in slight to general agreement, the fishing community neutral to slight disagreement (table 18).

TABLE 16. Tenure patterns of the upland farming community in the Maria River basin.

Farmland		Tenancy[a]		Pastureland[b]	
Owned	31	50:50	1	Owned	45
Tenanted	43	50:50[c]	2	Free grazing rights given by owner	1
Owned and tenanted	28	66:33	51	Communal grazing rights	54
Total	102	66:33[d]	2	Family arrangements	2
		75:25[d]	14	Total	102
		Other	1		
		Total	71		

Cattle	
Owned in their entirety	45
Shared equally between owner and tenant in an arrangement whereby the owner provides the cow, the tenant retains the first offspring , the owner the second offspring, and so on.	25
Shared equally, with the owner retaining the first offspring	2
CVRP program, whereby the first calf is retained by the tenant and the cow is passed on to the next recipient	2
Department of Agriculture program in which the farmer retains the cow and passes on the first calf to the next recipient	5
Department of Agriculture program whereby the farmer receives money to care for a breeder bull or bulls used in cattle dissemination	1
Family arrangements	1
N/A	18
Total	99

[a] Ratios represent sharing formula used in the crop harvest. The first value is the tenant's share, the second the landowner's. Ratios may differ slightly, depending on crop harvested.

[b] Applies to the grazing of all forms of livestock.

[c] Owner buys the seed and fertilizer.

[d] Tenant buys the fertilizer.

When questioned about recent trends in their specific livelihood sectors, discernible shifts toward pessimism occurred in all groups, but most starkly in the three coastal livelihood groups (table 18). However, it was the response to an ancillary question that underscored the gravity of the situation in the upland farming community. When asked if they wished their children to continue in farming, over two-thirds of the farmers responded in the affirmative (table 19). However when questioned why, what appeared to have been a positive response became one of pessimism: over half of the respondents stated that it was due to lack of options.

A subsequent set of questions attempted to explore in greater detail what factors were contributing to the range of perceptions regarding trends in the two communities' livelihoods (tables 20-23). Among the farmers who felt their lot was improving, government interventions were identified as chiefly responsible, while individuals feeling that productivity was

TABLE 17. Livelihood classification among coastal resource managers in the Maria River basin.

Occupation	Number employed, without additional means of support	Number so employed, with additional means of support[a]
Paddy farmer (a)	1	5
Coconut producer (b)	1	18
Fisherman (c)	1	2
Farmer (d)	2	16
(a) and (b)	1	11
(a) and (c)	2	1
(a) and (d)	3	6
(b) and (c)	0	3
(b) and (d)	2	10
(a), (b), and (c)	3	7
(a), (c), and (d)	3	1
(a), (b), and (d)	8	9
(b), (c), and (d)	0	2
(a), (b), (c), and (d)	1	1
Totals	28	92

[a] The most prevalent external means of support came from domestic or overseas remittances (21); from keeping livestock (17); and from working as laborers (16).

declining cited unproductive lands and/or drought as their biggest problems. When asked to rank issues affecting their livelihood negatively, farmers in a weighted analysis identified drought, fertilizer constraints, and unproductive lands/erosion as their three most critical problems.

Among the coastal community factors contributing to improvement in the paddy sector were new rice varieties, soil amendments, and irrigation. Factors contributing to decreases included pests, drought, and poor soils. Flooding and siltation were identified as issues, though neither was as prevalent as the above.[17] Drought and seasonal storms were the predominant issues cited by the coconut producers. Erosion generally and river bank erosion specifically followed in numerical order of responses. Finally, members of the fishing community attributed most of their problems to the presence of too many fishermen and/or illegal fishing.

A ranking of issues by paddy farmers identified pests, followed by soil fertility and water supply (apparently related to the drought), as the three of greatest significance. Problems of flooding/sedimentation, and soil erosion

---

17 In a recent survey of paddy farmers from *barangay* Lo-oc major issues identified were pests, disease infestation, and labor costs. See Antonio D. Achay, "A Management Survey of Twenty Five Rice Farms in Barangay Lo-oc, Maria Siquijor" (B.A. thesis, Foundation University, 1980), pp. 18-19.

TABLE 18. Resource managers' perceptions regarding their general economic situation and specific livelihood conditions.

	Relative agreement of respondents							
	StD	D	SlD	N	SlA	A	StA	Total
From the time you began growing paddy (harvesting coconuts; fishing; farming your land) in Maria, would you say your general economic situation has improved for you and your family?								
Paddy farmer	7	3	0	27	11	13	3	64
Coconut producer	8	6	1	29	19	12	1	76
Fisherman	0	2	1	13	6	3	1	26
Upland farmers	0	8	59	20	13	2	0	102
Since you first began growing paddy (harvesting coconuts; fishing; farming) in Maria, would you say your paddy production (trees; fish catch; present farm conditions) is increasing (are becoming more productive; has/have improved)?								
Paddy farmer	5	23	1	20	11	3	1	64
Coconut producer	10	28	0	25	4	1	0	68
Fisherman	0	15	5	4	0	0	0	24
Upland farmers	0	16	62	10	9	5	0	102

Note: No statistically significant differences were observed in responses among managers surveyed in the three coastal barangays (applying the Kruskal-Wallis test at the .1 significance level).

Key:
StD strongly disagree
D disagree
SlD slightly disagree
N neutral
SlA slightly agree
A agree
StA strongly agree

TABLE 19. Farmer interest in future generations' involvement in farming in the Maria River basin.

Farmers' wishes for their offspring	
Remain in farming	67
Leave farming	31
N/A	4
Total	102

Rationales for offspring remaining in or leaving farming

Remaining:		Leaving:	
No other options	57	Better employment alternatives elsewhere	13
Provides a good livelihood	2	Offspring dissatisfied with farming	6
Farmer needs labor inputs	8	Offspring already working elsewhere	10
		Total	96

TABLE 20. Perceived factors contributing to changes in livelihood productivity of upland resource managers.

Increases		Decreases		Don't know	Total
Government interventions	5	Pests (a)	3	4	102
Fertilizer	2	Drought (b)	18		
New technologies	1	Topographic constraints	1		
Other	5	Increasing labor costs	1		
		Fertilizer costs (c)	0		
		Unproductive lands (d)	22		
		Farmer incapacitated	1		
		(a) + (b)	4		
		(b) + (c)	6		
		(b) + (d)	10		
		(c) + (d)	7		
		Other	12		

were ranked fourth and eighth respectively. Among coconut producers it was drought followed by reduced production and soil erosion. River bank erosion was listed sixth. The responses among fishermen in ranking issues were at variance with numerical responses cited above, listing decreased catch first (although decreased catch is an outcome of the presence of too many fishermen, which was the issue ranked second). Sedimentation was ranked fifth behind equipment expense and illegal fishing.

The farmers' pessimistic views with regard to trends in both their general economic situation and their livelihood were paralleled by their demonstrated consensus in responses (80 to 90 percent) to several questions framed by the erosion issue. Specifically, responses to trends in soil loss, declining productivity of soils, and increased incidence of flooding indicated a worsening over time (table 24). When coastal users were asked similar, sediment-related questions, a division appeared among the paddy farmers with regard to both flooding and siltation (table 24). This contrasted with the range of responses among the coconut producers with regard to bank erosion. Most fishermen had no opinion.

For purposes of exploring varying perceptions, positive and negative, with regard to the significance of sediment to livelihood well-being, a series of questions were posed to the paddy farmers and fishermen. Most respondents in both groups viewed it negatively (table 25). However, to their credit, some paddy farmers identified sediment as a mixed blessing. In contrast, the fishing community's view was that sediment was detrimental to fish catch, although they were ambivalent with respect to its effects on coral reefs. This appears to be explained by their doubt as to whether coral reefs are living organisms (table 25).

TABLE 21. Perceived factors contributing to changes in livelihood productivity of coastal resource managers.

Increases		Decreases		Stayed the same	Don't know
Paddy farmers (n = 64)					
New rice varieties (a)	1	Pests (a)	4	7	3
Soil amendments (b)	3	Drought (b)	2		
Irrigation + (a) and (b)	7	Poor soils (c)	3		
Other	4	Flooding/siltation (d)	4		
		Overuse of soils (e)	2		
		(a) and (b)	3		
		(c) and (e)	3		
		(a), (b), and (c)	3		
		(a), (b), and (d)	3		
		Other	12		
Coconut producers (n = 75)					
Intercropping; new trees	1	Drought (a)	31	16	1
		Erosion (b) + (a)	3		
		Seasonal storms (c) + (a)	6		
		Eroding river banks + (a)	2		
		Poor soils + (a)	2		
		(a), (b), and (c)	2		
		Other	11		
Fishermen (n = 25)					
Too many fishermen (a)	8		0	2	3
Illegal fishing gear (b)	1				
Inappropriate fishing gear	2				
Legal constraints	2				
(a) + (b)	5				
Other	2				

Note: Data set did not meet chi-square test criteria.

Though data did not meet chi-square test criteria, responses among Lo-oc respondents did not appear discernibly different from other communities where CVRP has had less influence in the community through community education activities.

Watershed Dynamics. Another set of questions was posed to assess the degree to which a common understanding of physical processes contributing to erosion and sedimentation exists among the two communities. When asked to identify the predominant cause of erosion, the upland community responded with overutilized lands and/or rains, with the latter and/or flooding the chief means of transport (table 26). While a range of alternate routes were identified, almost all respondents named the coast as the final destination of eroded sediment.

The coastal communities demonstrated a similar understanding of the physical processes leading to downstream sedimentation. The two principal physical causes identified were rain and flooding contributing to erosion in the interior mountains, foothills, and/or adjacent farmlands (table 27). Climatic factors, topography, and erosion were among a number of factors believed to contribute to sediment transport.

Responsibility for Degradation. When asked what human actions were exacerbating soil erosion, there was a broad range of responses among farmers, of which the most prevalent were deforestation, apathy and carelessness, and ignorance (table 28). Overgrazing, as measured by number of responses, was not viewed as significant. Asked who is affected by sediment, the farmers were divided between thinking that it affected no one (either beneficially or adversely) and thinking that it affected only themselves.

The coastal livelihood groups showed consensus identifying the practice of *kaingin*-ing as the chief human source of erosion, followed by a range of physical factors. The coastal community demonstrated a pattern indicating that no one was to blame (because it was either God's will or due to physical factors beyond their control) or putting the responsibility on the farmers (table 29).

Both groups believed it was the combined responsibility of the government and farmer to resolve the issue (tables 28, 29).

Adjustments

If sedimentation is a critical issue, the adoption of mitigating measures should be correlated with community perceptions, barring the influence of constraining factors (lack of knowledge of adjustments, expense, etc.). Clearly, mechanical structures were the best known (table 30). These were most widely employed by the paddy farmers. Major constraints on adjustment were lack of government/community support and knowledge. In the farming community the most visibly employed device to prevent erosion is the rock wall terrace. Most farmers have three to five of these structures on their land, many built as far back as the 1920s by family members (table 31).[18]

---

[18] These structures serve to trap water-borne sediment and are planted in one or more of the region's common annual crops. Similar structures have been described elsewhere for use in areas of high local relief. In-filling appears to occur through a process described as seil irrigation (from the Negev) where soil surfaces exposed to the influence of rain and sheet-wash erosion increase transport which in turn rapidly in-fill behind the terrace. This appears to be

Paddy farmers were questioned about the cleaning of diversion canals, construction of flood works, and selection of rice varieties. While increased cleaning of canals was an adjustment resorted to (with the exception of Cantaroc B), defense works were not (table 32). A clear trend in the adoption of flood-resistant rice varieties occurred over time, from Catursa and Cainti some fifteen to twenty years ago to the newer hybrids IR-36 and IR-50 at present. However, it appears that the motives for these shifts were to obtain higher production and increased pest and drought resistance rather than to reduce losses attributed to flooding or sedimentation (table 33).[19] From the proposed adjustments for fishermen, increased time at sea and traveling greater distances were selected with greater frequency than switching gear and fisheries (table 34).

Critical Social Pressures. The results derived from comparing changes in farm ownership in the community of Cang-apa over a fifty-six-year period indicated that consolidation and parcelization were not important factors contributing to existing land-use patterns (table 35). During this period, owners of individual land parcels increased from 80 to 151. However, the numbers of single and multiple parcel owners also increased (approximately doubled), the latter dominated by individuals owning two or three parcels. As a result, a comparison of the respective owners:parcel ratios indicates virtually no change.

Similarly, responses to one survey question addressing changes in owner:tenant ratios over time indicated that increased tenancy was not a significant trend in Siquijor's upland areas (table 36). However, fragmentation was significant in Cang-apa as average parcel size decreased by approximately half over the period of comparison (table 35 and figure 13).

To better understand the characteristics and dynamics of land fragmentation in Maria's upland areas a set of questions attempted to determine how farmers obtained ownership rights to land and cattle. With respect to landownership, over two-thirds inherited their lands and another

---

the predominant process in Siquijor, as manual in-filling is constrained by thin soil mantles in the region and limited availability of local materials. The rock wall terraces also share the characteristics of Spencer and Hale's type 2 terrace, described as a high structure built in a narrow drainage channel in areas marked by high relief. The terrace consists of a massive stone barrage across a channel, which causes silting behind it, resulting in the production of a small, nearly horizontal field and is periodically soaked by naturally flowing water. See J.E. Spencer and G.A. Hale, "The Origin, Nature and Distribution of Agricultural Terracing," *Pacific Viewpoint* 2 (March 1961): 5-20.

[19] Catursa and Cainti are varieties from the Philippines. There appears to be very poor documentation with regard to their environmental characteristics. Catursa is an upland type while Cainti is a rather short and small-grained variety, neither in all probability particularly well-adapted to flooding and sedimentation. Personal correspondence from Dr. T.T. Chang, International Rice Research Institute, Los Banos, Philippines, 9 November 1988.

22 percent obtained their farms through purchase (table 37). Among those inheriting their lands, distribution followed one of two patterns: either lands were partitioned informally or a rotational system of sharing land was established.[20] It is interesting that despite the widespread use of rotational sharing of lands, owners preferred division of lands for their heirs by a ratio of 2:1. A similar desire can be discerned for the future disposition of cattle.

The documentation of birth and emigration patterns stemming from five generations of one family provided some insight into the role population increase plays in increasing pressure and demands on the land (table 38). In the subject family, emigration did not appear to occur until the third generation of record, whereupon half of all offspring left the community, the majority emigrating to Mindanao. Subsequently, the remaining offspring produced eighty progeny, resulting in nearly an equal rate of emigration but with a more diverse range of destinations. The remaining members of the fourth generation produced a much smaller cohort (forty-two, marking the fifth generation) than previous generations. This, it is suggested, is due to many of the fourth generation not yet having reached a childbearing age, and possibly to an increased awareness and use of methods to reduce family size. Similarly, the low rate of emigration is attributable to their relative youth. However, despite the role of emigration resulting in departures among approximately 50 percent of the offspring the cumulative totals of living individuals remaining on the land from the third, fourth, and fifth generations is significant.

Policy Effects

A final set of questions addressed the role of government programs in contributing to or mitigating erosion and sedimentation. Few farmers received assistance through the government's cattle disbursement or pasture improvement programs (table 39). Thus, the two national programs appeared to be of little consequence in contributing to or mitigating erosion in Siquijor. Similarly, in light of the intense land use, large number of owners and tenants, and few absentee landowners, the proposed policy regarding distribution of abandoned lands appears to be of little relevance to the island province. In terms of the farmers' perceptions regarding what the government could do, structural measures and training and education were preferred decidedly.

A similar absence of government response was perceived among the coastal community with the exception of the paddy farmers, some of whom

---

20 Typically this entails rotating access to the lands among siblings on a cropping or annual cycle with proceeds retained by the immediate user.

TABLE 22. Ranking of issues affecting upland livelihood well-being.

Rank	Power constraints	Fertilizer constraints	Drought	Unproductive lands/erosion	Pests	Seasonal storms	Stray animals	Irrigation constraints	Number of respondents[a]
1	10 (3) 30	29 (3) 87	34 (3) 102	14 (3) 42	4 (3) 12	4 (3) 12	0 (3) 0	2 (3) 6	97
2	4 (2) 8	21 (2) 42	26 (2) 52	28 (2) 56	9 (2) 18	4 (2) 8	3 (2) 6	2 (2) 4	97
3	3 (1) 3	6 (1) 6	7 (1) 7	7 (1) 7	11 (1) 11	6 (1) 6	0 (1) 0	0 (1) 0	40
Totals	41	135	161	105	41	26	6	10	
Rank[b]	4	2	1	3	4	6	8	7	

[a] Represents the total number of responses attributed to the respective rank.

[b] Ranking of issues was weighted assigning values of 3, 2, and 1 points to the first, second, and third priority issues respectively. The first figure under an issue represents the number of respondants, the second figure (in parentheses) the assigned weight, the third figure the total relative value for the issue (the product of the two preceding values).

TABLE 23. Ranking of issues affecting coastal livelihood well-being.

Rank	Water supply	Pests	Fertilizer cost	Soil fertility	Flooding/ sedimentation	Soil Erosion	Reduced productivity	Other	Number of respondents[a]
Paddy farmers									
1	11 (3) 33	10 (3) 30	4 (3) 12	11 (3) 33	5 (3) 15	2 (3) 6	2 (3) 6	7 (3) 21	52
2	4 (2) 8	9 (2) 18	0 (2) 0	6 (2) 12	1 (2) 2	0 (2) 0	3 (2) 6	4 (2) 8	27
3	5 (1) 5	3 (1) 3	4 (1) 4	2 (1) 2	5 (1) 5	1 (1) 1	2 (1) 2	2 (1) 2	24
Totals	46	51	16	47	22	7	14	31	
Rank[b]	3	1	6	2	5	8	7	4	

(Continued)

(TABLE 23 *continued*)

Rank	Water supply	Pests	Seasonal storms	Soil erosion	Reduced productivity	River bank erosion	Stealing	Labor constraints	Number of respondents[a]
Coconut producers									
1	41 (3) 123	3 (3) 9	4 (3) 12	6 (3) 18	6 (3) 18	2 (3) 6	2 (3) 6	2 (3) 6	66
2	9 (2) 18	5 (2) 10	4 (2) 8	6 (2) 12	3 (2) 6	1 (2) 1	0 (2) 0	0 (2) 0	28
3	3 (1) 3	0 (1) 0	8 (1) 8	1 (1) 1	8 (1) 8	2 (1) 2	0 (1) 0	0 (0) 0	22
Totals	144	19	28	31	32	10	6	6	
Rank[b]	1	5	4	3	2	6	7	7	

Note: Data set did not meet chi-square test criteria.

Rank	Too many fishermen	Decreasing catch	Equipment expense	Illegal fishing	Maria Bay sedimentation	Legal constraints	Number of respondents[a]
Fishermen							
1	9 (3) 27	8 (3) 24	3 (3) 9	3 (3) 9	0 (3) 0	1 (3) 3	24
2	3 (2) 6	5 (2) 10	1 (2) 2	0 (2) 0	2 (2) 4	0 (2) 0	11
3	0 (1) 0	0 (1) 0	0 (1) 0	0 (1) 0	0 (1) 0	1 (1) 1	1
Totals	33	34	11	9	4	4	
Rank[b]	2	1	3	4	5	5	

[a] Represents the total number of responses attributed to the respective rank.

[b] Ranking of issues was weighted assigning values of 3, 2, and 1 points to the first, second, and third priority issues respectively. The first figure under an issue represents the number of respondents, the second figure (in parentheses) the assigned weight, the third figure the total relative value for the issue (the product of the two preceding values.)

TABLE 24. Upland and coastal resource managers' perceptions of selected environmental issues.

	Relative agreement of respondents						
	StD	D	SlD	N	SlA	A	StA
Upland farmers (n = 102)							
Loss of soil from rain and runoff is getting worse	0	2	5	1	63	30	1
Soils are becoming less productive	0	3	13	0	42	43	1
Flooding of lands is getting worse	0	5	11	1	53	32	0
Paddy farmers (n = 64)							
Paddy fields are increasingly flooded	9	21	0	17	13	3	1
Fields are increasingly affected by siltation	9	21	0	17	14	2	1
Coconut producers (n = 36)[a]							
Trees are increasingly affected by erosion and the caving in of river banks	8	10	1	6	8	3	0
Fishermen (n = 25)							
Increasingly, river-borne sedimentation is killing the reefs	0	5	0	20	0	0	0

Note: No statistically significant differences were observed in responses among managers surveyed in the three coastal barangays (applying the Kruskal-Wallis test at the .1 significance level).

Key: StD strongly disagree; D disagree; SlD slightly disagree; N neutral; SlA slightly agree; A agree; StA strongly agree

[a] Results presented only from respondents harvesting coconut trees along the river's edge.

apparently received some assistance for construction of water control structures (table 40).

At the provincial level there was no evidence that export ceilings affected grazing densities one way or another. Interviews with both shippers and port officials indicated that bans limiting the export of livestock were rarely enforced and ineffective.

Despite existing legislation against the grazing of cattle on other private lands the practice is omnipresent. This appears, however, to be more a failure of the public to resort to legal mechanisms for relief than an example of institutional failure. One possible explanation for this is the widely held Filipino social value on maintaining smooth interpersonal relations and conformity. This is particularly evident in the country's rural areas where Western values have had less influence. A second possible explanation may

TABLE 25. Comparative perceptions of selected biophysical interactions in Maria River basin among coastal resource managers.

	Lo-oc	Cantaroc B	Cantaroc A	Total
		(number of respondents)		
Paddy farmers				
Effects of sediment on paddy fields				
Beneficial: provides natural fertilizer	3	1	3	7
Harmful: damages or destroys crop	16	9	12	37
Both beneficial and harmful	2	6	3	11
Neither beneficial nor harmful	2	0	4	6
Don't know	1	0	2	3
Total number of respondents	24	16	24	64
Fishermen				
Effects of sediment on fish catch				
Harmful: destroys habitat, ruins fishing gear, and reduces water quality	13	6	4	23
No effect	0	0	1	1
Don't know	0	0	0	0
Total number of respondents	13	6	5	24
Effects of sediment on coral reefs				
Harmful: smothers coral and impedes fish from laying eggs	4	0	0	4
No effect	4	3	0	7
Don't know	6	3	5	14
Total number of respondents	14	6	5	25
Coral reefs as living organisms				
Living	1	0	0	1
Nonliving	6	1	1	8
Don't know	7	5	4	16
Total number of respondents	14	6	5	25

Note: Data set did not meet chi square test criteria.

be a fatalistic belief that unimproved lands have become so unproductive over the years that additional grazing pressure is of little consequence. Support for the first suggestion is derived from the common response among farmers that open grazing is a tradition passed down from previous generations and a local custom. Support for the latter explanation revolves around the one stipulation in the law that found support in the survey, that is, the protection of land improvements from stray cattle.

TABLE 26. Perceptions of selected river basin processes among upland hilly land farmers.

Physical processes contributing to erosion		Physical factors contributing to sediment transport	
Overutilized lands (a)	7	Rain (a)	59
Rain-driven erosion (b)	54	Flood (b)	4
Drought[a] (c)	4	Typhoon	2
Overtillage (d)	3	Wind	2
(a) + (b)	11	(a) + (b)	26
(b) + (c)	5	Other	5
(c) + (d)	3	Don't know	4
Other	13	Total number of respondents	102
N/A	2		
Total number of respondents	102		
Destination of sediment			
Contours, canals, coast[a]	18	Paddy, river, coast	3
River coast	36	Adjacent farm, river, coast	2
Valley, lowlands, coast	14	Don't know	3
Coast	26	Total number of respondents	102

[a] Refers to soil conservation measures used in hilly land areas to reduce water runoff and soil loss.

TABLE 27. Perceptions of selected river basin processes among coastal livelihood groups.

	Paddy farmers	Coconut producers	Fishermen	Total no. of respondents
Physical processes contributing to sedimentation				
Rain-driven erosion: hills, rivers, lowlands	57	54	11	122
Rain-driven erosion: hills, rivers, lowlands, oceans	0	2	6	8
Flooding	4	7	4	15
Other	3	12	4	19
Total number of respondents	64	75	25	164[a]
Physical factors contributing to sediment transport				
Rain, wind, typhoons, and floods (a)	15	6	8	29
Soil erosion (b)	0	42	4	46
Physical topography + (a)	26	0	2	28
Physical topography + (b)	4	7	3	14
Other	19	20	8	47
Total number of respondents	64	75	25	164[a]
Source of sediment				
Mountains, hilly lands (a)	49	52	14	115
Adjacent farmlands (b)	14	0	1	15
(a) + (b)	1	23	3	27
Other	0	0	7	7
Total number of respondents	64	75	25	164[a]

[a] Data set did not meet chi-square test criteria.

TABLE 28. Upland farmers' perceptions of the human role in erosion.

Human factors contributing to erosion		Persons affected by sediment		Responsibility for issue resolution	
Overutilized soils (a)	5	No one	42	Farmer	21
Deforestation, slash/burn (b)	23	Respondent	51	Government	5
Overtillage (c)	2	Adjacent farmer	3	CVRP	0
Carelessness, laziness, apathy (d)	18	Other farmers	0	Farmer, CVRP	1
Overgrazing (e)	7	Other	6	Farmer, government	75
Ignorance (f)	5	Total respondents	102	Total respondents	102
(a) + (b)	2				
(b) + (d)	5				
(b) + (e)	2				
(d) + (f)	12				
Other	15				
Don't know	6				
Total respondents	102				

TABLE 29. Coastal livelihood groups' perceptions of the human role in erosion.

	Paddy farmers	Coconut producers	Fishermen	Total no. of respondents
Human factors contributing to sedimentation				
Slash and burn	19	41	8	68
Physical factors	9	22	4	35
Other	36	12	13	61
Total number of respondents	64	75	25	164[a]
Responsibility for sediment effects				
No one/God's will	4	8	4	16
No one/physical factors (a)	9	14	2	25
Upland farmers (b)	2	1	0	3
(a) + (b)	22	35	1	58
Other	27	17	18	62
Total number of respondents	64	75	25	164[b]
Responsibility for issue resolution				
No one/God's will	8	19	3	30
Government/upland farmers	40	44	15	99
Other	16	12	7	35
Total number of respondents	64	75	25	164[c]

[a] No significant differences occurred among barangay responses at either the .05 or .1 levels of significance.
[b] Data set did not meet chi-square test criteria.
[c] Significant differences occurred among *barangay* responses at .1 and .05 levels of significance.

TABLE 30. Knowledge, use, and constraints to choice of mitigative adjustment among coastal resource managers to upland-derived resource use conflicts.

	Paddy farmers	Coconut producers	Fishermen
What measures exist to prevent or avoid the flooding and siltation of your paddy (loss of coconut trees adjacent to eroding banks; effects of dirtying Maria Bay)?			
Diversion canals, rock wall terraces, dikes, and fencing (a)	33	35	7
Reforestation (b)	1	3	5
(a) + (b)	11	6	5
Intercropping, strip cropping, and no tillage	1	3	0
Other (mainly combinations of existing categories)	9	10	3
None	5	12	1
Don't know	4	6	4
Total number of respondents	64	75	25
Which of these measures do you use?			
Diversion canals, rock wall terraces, dikes, and fencing (a)	34	5	5
Reforestation (b)	0	1	0
(a) + (b)	0	0	0
Intercropping, strip cropping, and no tillage	0	3	0
Other (combinations of existing categories)	2	7	1
None	26	58	19
Don't know	0	0	0
Total number of respondents	62	74	25
Why not use the others you list?			
Labor constraint	0	2	2
Financial constraint	2	3	1
Time constraint	2	2	0
Lack of government or community support	6	4	4
Lack of knowledge	4	6	0
Not his land	1	3	1
N/A	43	51	18
Total number of respondents	63	71	26

Note: Data set did not meet chi-square criteria.

TABLE 31. Characteristics of stone wall terraces as soil conservation structures in Maria River basin, Siquijor Province.

Number of terraces per farming unit		Period of terrace construction		Individuals responsible for construction	
0 terraces	18 farms	1920-30	12	Farmer	31
1-2	21	1931-40	0	Father	12
3-5	31	1941-50	6	Grandparents	6
6-10	28	1951-60	11	Great-grandparents	6
11-20	4	1961-70	8	Farmer association	9
Total	102	1971-80	7	In-laws	3
		1981-present	37	Laborers	1
		Don't know	3	Other	16
		Total	84	Total	84

TABLE 32. Selected adjustments to upland-derived resource use conflicts among paddy farmers.

	Yes	No	Total
From the time you began growing paddy, have you had to increase the frequency of cleaning your water diversion canals?			
Lo-oc	18	6	24
Cantaroc B	7	9	16
Cantaroc A	18	6	24
Total number of respondents	43	21	64[a]
From the time you began growing paddy, have you had to build defense works to protect your fields from flooding?			
Lo-oc	12	12	24
Cantaroc B	2	14	16
Cantaroc A	6	18	24
Total number of respondents	20	44	64[b]

[a] Application of the chi-square test indicated a difference in responses among *barangays* at the .05 significance level.

[b] No significant difference occurred among *barangay* responses at either the .1 or .05 levels of significance.

TABLE 33. Trends and rationales in usage of rice varieties among paddy farmers in Maria River basin, Siquijor Province.

Paddy farmer usage of specific varieties of rice over time					Rationales for usage of present rice variety	
Varietal	Present	5 yrs ago	10 yrs ago	15 yrs ago	Rationale	No. of respondents
IR-50	7	14	10	0	Higher production (a)	34
IR-36	42	40	10	0	Pest resistance (b)	0
BPI	5	6	4	0	Drought resistance (c)	0
Catursa	0	0	16	32	(a) + (b)	9
Cainti	0	0	17	35	(a) + (b) + (c)	5
Burikat	3	3	0	4	Other[b]	16
Ceres	1	4	0	0	Total no. of respondents	64
Others	3	2	5	2		
Totals[a]	61	69	62	73		

[a] Table includes multiple responses indicating concurrent use of more than one variety during respective time periods.

[b] Wind resistance, low fertilizer demand, government recommendations, and peer usage.

TABLE 34. Selected adjustments to upland-derived resource use conflicts among fishermen.

	Yes	No	Total
From the time you first began fishing, have you had to increase the total daily time spent to catch the same amount of fish?			
Lo-oc	8	5	13
Cantaroc B	5	1	6
Cantaroc A	3	2	5
Total number of respondents	16	8	24
From the time you first began fishing, have you had to go further to catch the same amount of fish?			
Lo-oc	9	5	14
Cantaroc B	1	5	6
Cantaroc A	2	3	5
Total number of respondents	12	13	25
From the time you first began fishing, have you had to switch to different fishing gear to catch other species of fish, owing to declining yields of your preferred species?			
Lo-oc	6	8	14
Cantaroc B	1	5	6
Cantaroc A	2	3	5
Total number of respondents	9	16	25

Note: Application of the chi-square test to measure differences in responses was not possible, owing to failure to meet test conditions.

TABLE 35. Changes in patterns of farm parcel ownership and mean parcel size, 1927-83, *barangay* Cang-apa, Province of Siquijor.

		Number of owners	
		1927	1983
Number of parcels:	1	56	108
	2	15	28
	3	4	8
	4	3	4
	5	1	2
	6	1	0
	7	0	0
	17	0	1
Total single-parcel owners		56	108
Total multiple-parcel owners		24	43
Subtotal[a]		80	151
Other owners[b]		19	22
Total owners		99	173
Ratio of owners : parcels		.66	.65
Average number of parcels per owner[c]		1.51	1.49
Mean parcel size (ha)[d]		2.08	1.10

[a] Application of the chi-square test indicated no significant differences between the two periods in terms of distribution of parcel ownership at the .05 level.

[b] "Other" designates public lands and parcels classified in the generic category of "heirs" (of the former owners) without further specificity.

[c] These figures exclude the categories of parcels designated as public lands and lands belonging to heirs.

[d] The summation of individual parcel sizes taken from tax records provided an estimated total area of 181.91 ha for the *barangay* of Cang-apa. Tabular mean values have not discounted for the heirs and public lands categories. To factor these out the total estimated area would be reduced to 154.72 and 167.86 ha for the years 1927 and 1983 respectively. This in turn reduces the means to 1.93 and 1.11 ha for the same years.

TABLE 36. Temporal changes in farmland tenure patterns in the Maria River basin's upland farming community.

Changes from landowners to tenant farmers	
Owner (no change)	59
Tenant (no change)	40
Former owner/present tenant (change)	3
Total	102

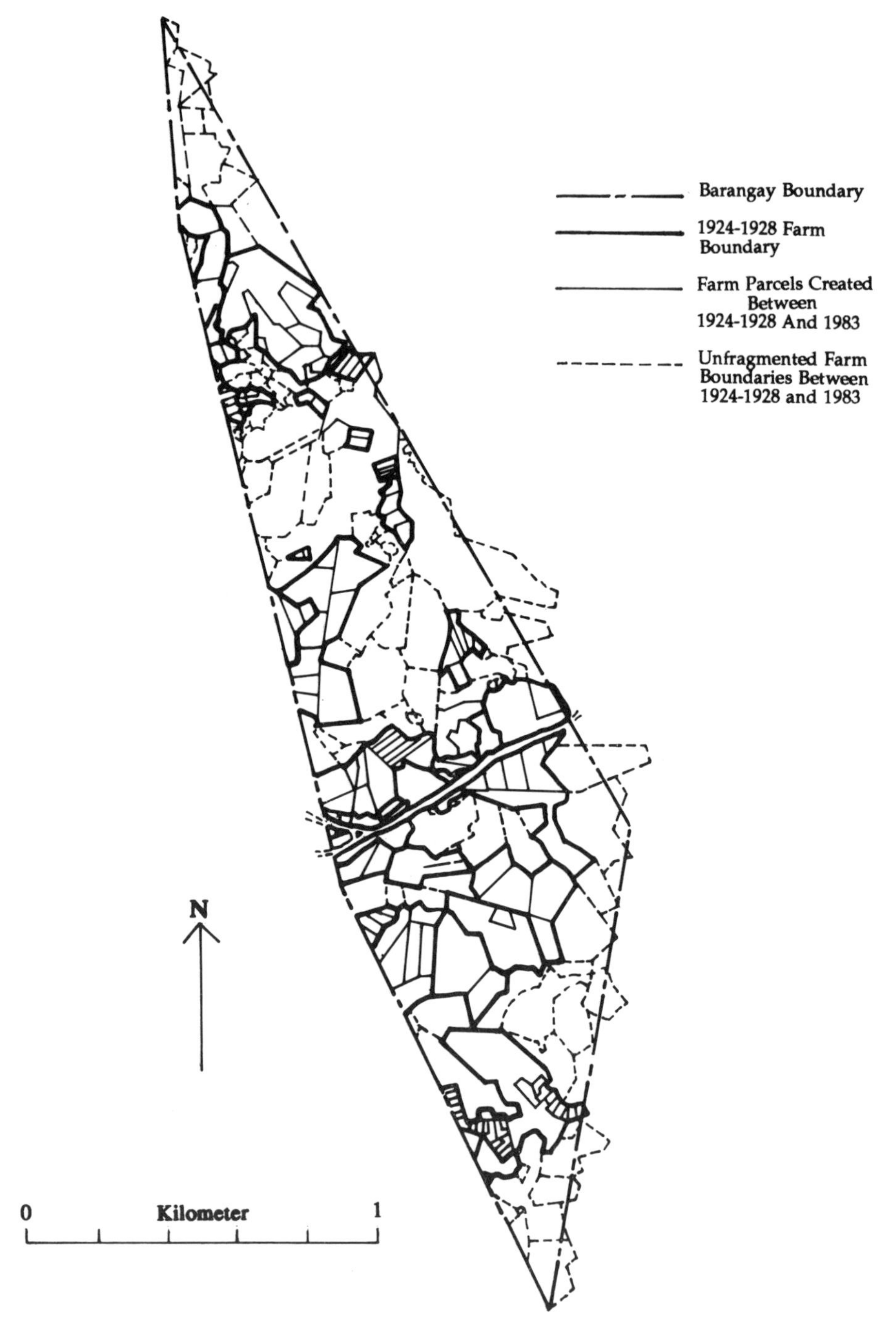

Fig. 13. Land fragmentation, barangay *of Cang-apa*

TABLE 37. Patterns of landownership in the upland farming community in the Maria River basin.

Source of ownership		Farmer inheritance pattern	
Purchased	14	Sole heir	3
Inherited	45	Equal sibling sharing[a]	17
Mortgaged	1	Rotational sharing	16
Through marriage	2	Land partition	1
Other	2	Sibling abandonment of rights	2
Total	64	Other familiar arrangements	6
		Total	45
Heir inheritance pattern		**Cattle inheritance by heirs**	
Siblings divide	7	Siblings divide equally	36
Parent divides equally	33	Cattle sold, money divided	7
Land held communally in family	14	Cattle rotated among heirs	10
Eldest son divides equally	2	Siblings decide	7
Doesn't know	2	Doesn't know	3
N/A[b]	44	Other	3
Total	102	N/A[c]	36
		Total	102

[a] This is effectively land partitioning less the legal process. Legal partitioning is rarely resorted to, owing to the belief that additional expenses associated with registration and taxes will be incurred. The shared characteristic between the two approaches is the right to farm the land and keep the fruits of the respective sibling's burden.

[b] Refers to individuals with no land and/or no children.

[c] Refers to individuals with no cattle and/or no children.

TABLE 38. Population growth and emigration characteristics in five generations of the Calibo family, Barangay Cang-apa, Province of Siquijor.

Population characteristics

Generation	Period[a]	Births, deaths	Number of people	Migrants	Remained	Died[b]	N/A[c]	Offspring	Still Alive	Average age (years)
1	1830-1889[d]	0, 0	-	-	1	-	-	7	-	-
2	1860-1950	1, 1	7	-	7	-	-	44	-	-
3	1921-1978	35, 8	44	27	14	1	2	80	6	66
4	1953-1968	59, 2	80	37	36	1	6	42	33	41
5	1975-	30, 0	42	4	31	4	3	-	31	13

Migratory characteristics

	Destination							Sex		
Generation	Siquijor	Central Visayas	Mindanao	Manila	United States	Other	Year of migration (mean)	Male	Female	Total
3	6	-	18	1	2	-	1952	13	14	27
4	14	6	8	1	6	2	1970	15	22	37
5	1	-	3	-	-	-	?	3	1	4

[a] Calculated from mean year of births and deaths respectively.
[b] Those under the age of fifteen remaining in Cang-apa.
[c] Those about whom no information was obtained.
[d] Estimated from oral histories.

TABLE 39. Upland institutionally sponsored mitigative measures to reduce downstream resource use conflicts.

Government assistance directed to cattle disbursement[a]			
Cattle received:		Other government assistance:	
None	7	None	26
1	7	Training, education	17
N/A	88	Planting materials	3
Total number of respondents	102	Veterinary services	7
		N/A	49
		Total number of respondents	102
Government assistance directed to pasture improvement[b]			
None	23	Other	1
Training, education	3	N/A	45
Planting materials	30	Total number of respondents	102
Soil loss reduction and fertility improvement			
Government assistance:		Government assistance desired by respondents:	
Training, education	3	Training, education	13
Planting materials	3	Planting materials	6
Structural measures	12	Structural measures	36
Soil amendments	8	Structural measures and training	29
Other[c]	25	Other	5
None	3	Don't know	11
N/A	48	N/A	2
Total number of respondents	102	Total number of respondents	102

[a] Of the 102 respondents, 14 participated in the Department of Agriculture/Bureau of Animal Industry Cattle Disbursement Program.

[b] Of the 102 respondents, 5 participated in the Department of Agriculture/Bureau of Animal Industry Pasture Improvement Program.

[c] Previously designated categories in combination with structural measures.

TABLE 40. Institutionally sponsored mitigative measures to reduce flooding and sedimentation on the coast.

	Paddy farmers	Coconut producers[a]	Fishermen
River canalization	13	0	0
Other[b]	12	0	6
None	39	75	19
Total number of respondents	64	75	25

Note: Data set did not meet chi-square criteria.

[a] The question directed to coconut producers addressed the issue of riverbank erosion.

[b] Dams and dikes, seminars, and soil amendments for paddy farmers; artificial reefs for fishermen.

Chapter 6

ANALYSIS, DISCUSSION, AND CONCLUSION

Analysis

The Biophysical Component

The results from the resource assessment and one-year monitoring program in the Maria River basin indicate an area extremely impoverished in terms of natural resources, a condition that appears more a result of human pressures accumulating over time than endowment constraints.

The principal physical factors contributing to the omnipresent state of severe erosion in Siquijor's highlands are regional precipitation patterns in combination with the island's physiographic characteristics. Although empirical data are limited, rainfall patterns indicate a monsoonal regime marked by frequent drought, the occurrence of which was observed in the present study. Precipitation during the southwest monsoon seems best described as frequent but limited in duration and areal scope, a pattern broken only by the occasional, high-intensity storm capable of affecting a much broader area. In contrast, rainfall during the northeast monsoon is relatively insignificant. The rain's effect on surface erosion is exacerbated by the absence of natural vegetation, steep slopes, and shallow and highly erodible soils whose low permeability contributes to rapid surface runoff.

Although short-duration rainfalls undoubtedly contribute to upland erosion, the absence of discernible flood peaks at the Maria River mouth monitoring station indicates that the few, high-intensity events are the major source of sediment transport affecting the coast.

Sediment can perhaps be best described as reaching the coast in a series of pulses carried by occasional rainfall and surface runoff, a journey by no means continuous, but likely to occur in alternating cycles of deposition and resuspension. At the coast, a new energy regime dominated by nearshore currents and/or wind-driven swell during the northeast monsoon replaces the watershed's drainage processes. Under southwest monsoon conditions sediment appears most likely to settle out in immediate proximity to the river mouth, the exceptions being during long-period ebb tides and/or strong southerly wind and swell conditions, whereupon nearshore currents may carry sediment to the northernmost portions of the bay. With the onset of the northeast monsoon, a situation characterized by dramatically reduced rainfall and the formation of a sandbar at the river's mouth, sediment inputs decline dramatically.

Over time, these processes have resulted in a number of downstream morphological modifications. Onshore these include the formation of a fertile, albeit narrow, coastal flood plain in some places and severe river bank erosion attributed to the effect of river-borne sediment. Offshore, the principal feature is the mud-dominated substrate occurring in the bay's central and southern portions beyond which the bottom becomes increasingly sandy; in the bay's northern and southern sectors, it is characterized by the presence of small patch reefs. The absence of mud bottom in waters to the north of the Maria River is best explained by the lack of well-defined surface drainage in the area, in contrast to the south.

The existence and magnitude of the bay's mud bottom is significant for a number of reasons. First, its presence, continuously fed from terrigenous sources, is a major factor in preventing the establishment of highly diverse and productive benthic communities such as coral reefs and marine grassbeds. Second, existing corals, particularly those in proximity to the mud zone, appear to be suffering the effects of sediment-related stress. However it must be noted that there also existed sporadic evidence of destructive fishing methods, resulting in the degradation or death of corals. In those instances it was impossible to distinguish whether death resulted from fishing methods or the effect of sedimentation. Finally, the mud bottom appears to be the major source contributing to bay turbidity during the northeast monsoon, a result of the monsoon's characteristic easterly driven swell, which provides the energy to force surficial muds into resuspension. The near-chronic turbidity associated with the continued pounding of incoming swell contributes to prolonged periods of sediment resuspension in the water column, providing ample time for tidal forces to exert their maximum influence on current direction and velocity and resulting in an expanded zone of stress. This latter process is perhaps more significant in affecting water quality (and thus responsible for its attendant effects on some living benthic communities) than the infrequent sediment-laden runoff associated with the southeast monsoon. This view is supported by monthly visibility averages indicating less turbidity in the bay's northern waters throughout the year, a pattern explained both by the aforementioned absence of drainage and the sheltering influence from offshore swell provided by the bay's rocky headland and, in all likelihood, is one factor explaining the increased living coral cover found in these waters relative to the central and south-central portions of the study area.

Surface salinity, while demonstrating changes in monthly values ranging from 29 to 35 ppt, did not vary significantly among stations in any one month and is regarded as inconsequential in explaining spatial differences in biotic distribution and production.

The Socioeconomic Component

The study's findings with regard to both upland and coastal land-use practices indicate intense pressure on the island's land and marine resources, supporting the conclusion that both communities are living at the margin in terms of natural resource–based economic dependency. In the upland areas, land-use practices commonly considered nonsustainable that appear to be contributing significantly *in toto* to the problem of erosion include monocropping, triple-cropping of cornfields, failure to incorporate a fallow period into the production cycle, and overgrazing. In the coastal lowlands most resource users appear to depend heavily on multiple livelihoods and offshore remittances to supplement their incomes.[1] Paddy farmers have increasingly turned to high-yield, short-term rice varieties and the use of fertilizers as a means to increase production. Offshore, fish standing stock appears relatively low and there is evidence of destructive fishing practices (such as the use of dynamite), both indications of an impoverished fishery.

Cattle grazing patterns and the status of pasture, the upland land use selected for the study's focus, reflect the aforementioned pattern of intensification indicated by high grazing densities (2.1 head of cattle and carabao combined/ha of unimproved pasture), a near-universal absence of improved pastures, and the widespread practice of uncontrolled grazing.[2]

The effects of erosion on selected coastal resource users' livelihoods attributable to the river basin's physical processes in combination with cur-

---

1 This is probably prevalent in the upland community as well, but no questions addressing these topics were included in the upland survey.

2 The 2.1 head/ha figure is derived from the results of the survey questionnaire. Density estimates obtained from the Cang-apa farm fragmentation study with its smaller average farm size indicated a density of 2.7 head/ha (this assumes pasture:farm size and herd size:farm ratios are identical to the survey results). Data from the University of San Carlos study indicated an average farm size of .36 ha and combined herd size of 4.3 head/farm combining to yield a density of 11.9 head/ha. This latter figure seems unrealistic for *Imperata* pasture based on reviews of documented grazing rates. Discrepancies in the former two figures could be explained by errors in farmers' estimates of areal farm size (survey) and unregistered land partitions, a practice believed to be prevalent in Siquijor (fragmentation study). For our purposes we will use the 2.1 cattle/carabao figure calculated from the survey. However even this most conservative of figures contrasts sharply with the existing government policy recommending that grazing density be limited to 1 head per 4 ha of unimproved, unamended grasslands. For a review of literature describing grazing trial results on *Imperata cylindrica* in Southeast Asia see Falvey, "*Imperata cylindrica.*"

rent land-use practices are readily apparent. These include severe river bank erosion and sediment-related stress to corals, which affect coastal users' economic well-being through loss of coconut trees and highly productive, sediment-sensitive fish habitat, respectively.[3] Within the study period no flooding or siltation was observed in the coastal paddy fields. However, oral histories appear to confirm the periodic flooding and siltation of these fields, which contribute to declines in paddy production and/or loss of rice crops.

The study was not able to differentiate the human role from the biophysical sources contributing to downstream sediment-related resource issues. In light of this constraint the resource users' relative ratings of environmental issues affecting their livelihoods was used as a surrogate in assessing relative importance. Employing this approach, erosion was viewed as one of the three most critical issues affecting upland farmers. In contrast, most members of the coastal community viewed sedimentation-related issues as relatively low priorities ranked behind drought and other sector-specific issues (e.g., pests, illegal fishing). A correlative effect of the drought was to reduce runoff and downstream effects attributed to waterborne sediment. The perceived importance of drought among the paddy farmers and coconut producers provides support for one hypothesis forwarded in early hazards research regarding the factors of recency and severity of hazard in influencing human perception of same.[4] Moreover, the coconut producers affected by bank erosion were limited to those owners with parcels located adjacent to the river bank. However, these factors fail to explain the response of the fishermen, who were more preoccupied with declines in fish production and increasing competition than sediment-related coral reef degradation. In part, this may be explained by the conceptual complexity required to associate coral reef health and fish production to the effects of sedimentation. The previously described ambiguity with respect to the nature of coral reefs provides some evidence for this explanation.

Although categories were not mutually exclusive, both upland and coastal communities demonstrated a high degree of understanding with respect to the physical forces affecting soil erosion and sedimentation, transport processes, and the origin and destination of eroded soil. However, their knowledge of the human forces contributing to the issue was more am-

---

3 Poor land management is widely recognized as leading to decreased production of reef communities and ultimately their destruction. Particularly well documented cases have been reported from Australia and Hawaii. See Ferguson Wood and Johannes, *Tropical Marine Pollution*, p. 17; J.E. Maragos, "A Study of the Ecology of Hawaiian Reef Corals" (PhD thesis, University of Hawaii, 1972); and R.W. Fairbridge and C. Teichert, "The Low Isles of the Great Barrier Reef: A New Analysis," *Geographical Journal* 111 (1948): 67-88.

4 See G.F. White, "Natural Hazards Research," in *Directions in Geography*, ed. R.J. Chorley (London: Methuen and Co., 1973), pp. 193-216.

biguous. One widely held belief among both groups was in the importance of shifting agriculture in contributing to erosion. While this practice may have been significant in leading to devegetation at one time, it appears of little consequence at present on an island largely devoid of any natural vegetation. Similarly, the view that soil loss is not a function of human activity but is primarily a manifestation of the forces of God or nature perhaps is indicative of a certain belief that may be characteristic of the culture. Finally, there is the failure among the farming community to associate the intensive overgrazing with the severe soil erosion characteristic of Siquijor's upland marginal lands. However, the substantial number of respondents from both communities identifying upland farmers as the group responsible for both the issue and its resolution is important, the significance being the common understanding of the problem's social dimensions.

Responses to a number of questions demonstrated no statistical differences among the three coastal communities. One implication of the number of statistically similar responses between coastal inhabitants within and outside of the Maria River basin boundaries is that perceptions of physical and social processes contributing to downstream sedimentation may not be unique to the single river basin. Similarly, it would also appear to confirm that CVRP-sponsored local education activities did not significantly influence the perceptions of Lo-oc residents, particularly with regard to the effect of sediment on coastal resources.

In the upland areas the most visible adjustment to chronic soil loss was the construction of stone wall terraces, a feature of the landscape that apparently extends back at least as far as the 1920s and most probably earlier.

Knowledge among coastal users of existing and potential adjustments to mitigate erosion was broad, albeit with a decided bias detected in responses toward mechanical solutions versus "soft" approaches (such as hedgerows and soil enrichment).

Only a few paddy farmers had taken precautions, structural or otherwise, to counter the risk of flood. However, increased cleaning of irrigation ditches had occurred in communities where paddy farmers were organized. Similarly, coconut farmers with lands adjacent to the river had taken no action to counter the effects of bank erosion and appeared resigned to its occurrence. Among members of the fishing community there had been an apparent increase in fishing effort but less evidence of traveling greater distances or changing fisheries, an adjustment that cannot be attributed solely to the negative effects of sediment on fish habitat.

One could argue that the low priority accorded sedimentation as hazard among the coastal community could be correlated with the low number of adjustments identified in the survey. However, the absence of adjustments beyond accepting risk perhaps can be better explained by one or more of the following hypotheses: specific resource-related adjustments may have been supplanted by other responses related to the broader economic environment (e.g., emigration); limited financial resources and a probable inverse correlation between willingness to accept hazard risk and availability of financial resources to do something about it; compounded adjustment (e.g., increased effort in response to overfishing may also serve as an effective response to sediment effects); and human factors (e.g., the survey's high incidence of responses citing laziness/apathy among upland farmers as a contributory cause to erosion).

The results of the study's historical analysis demonstrate the importance of both local and external economic forces in influencing the island's current land-use practices in upland areas. Although documentation is scarce, it appears that precolonial populations in the Central Visayas were hunter-gatherers largely concentrated along the islands' coasts. With the imposition of a new order by the Spanish, a number of changes were implemented over time which, to varying degrees, bore relevance to the present study. These included the imposition of a market economy over a more traditional form of exchange; the introduction of exotic crops for purposes of supporting colonial outposts and later providing a source of exports; the likely, though undocumented, role of the Spanish in logging and deforestation in the region; and the establishment of a top-down administrative structure.[5]

Until the end of the nineteenth century the island's exports were dominated by agricultural products such as tobacco and abacá, while local food consumption was dominated by corn and rice. It is unclear when the island lost its self-sufficiency in these basic staples, but there is evidence of the need to import substantial amounts of essential foodstuffs as early as the 1920s, a situation that has continued to the present.[6]

One assumes that Siquijor's loss of self-sufficiency together with declines in the island's world markets in abacá and tobacco or loss of its competitive advantage made for a precarious economic (and social) situation. Relief appears to have come with the onset of the American period and the creation of a number of new national markets, including one for beef. Market demand, perhaps influenced to varying degrees by the island's physical

---

[5] The cultivation of local plants and introduction of exotic plants in response to external markets was to exert a profound impact in the Philippines. One result was the establishment of large plantations on the country's fertile lowlands, a feature cited later as a major cause of landless farmers' encroachment on marginal lands. However it is doubtful whether this pattern applies to Siquijor, considering the island's mostly narrow coastal plain and the high percentage of titled ownership of highlands extending back to the turn of the century.

[6] A draft of the province's newest capital development plan indicates that both rice and corn production have continued to decline. See R.P., Siquijor Provincial Development Staff, *Capital Development Plan,* p. 6.

constraints, government policies with regard to the carabao, American agricultural technology, and growth and diversification of the usefulness of cattle as sources of power, food, and savings, provided much of the basis for the dramatic shifts observed in export products in the first thirty years of the twentieth century (though copra exports continued to grow in importance as an export).[7]

Despite the increasing importance of cattle in the island's export economy, it is apparent that the industry has not provided the benefits required to support the upland community's basic needs over time. In part this appears to be attributable to population growth and its effects on farm area and number of family members dependent on the farm. The shrinking size of farms indicated in the Cang-apa land fragmentation study appears to be adversely affecting the well-being of the cattle owner and the community at large. This may occur either through the loss of income associated with the decreases in herd size accompanying smaller pastures, or reduced productivity of land and cattle attributable to intensified grazing pressures resulting from maintaining present herd size on ever smaller farm lands. Evidence indicates that this latter process may be the more important of the two, occurring through an unusual variant on the "tragedy of the commons."[8] The upland farming community appears to support open grazing on private lands and the farmers themselves show considerable tolerance of stray cattle grazing their lands, rendering these private lands into a commons. As Hardin's model postulates, there are few impediments and no natural resource constraints associated with an ever decreasing farm size to prevent a new farmer from acquiring cattle and releasing them to graze the hillsides.

Besides fragmentation, population growth in the face of a fixed farm size appears to be contributing to the establishment of a number of informal familiar arrangements providing for access to limited land resources, of which the rotational sharing of land among siblings appears to be the most common. The significance of the latter lies not only in the chronic need of siblings to find alternative means of livelihood but in effect creates a queue of individuals waiting to work the land, thus causing additional resistance to incorporating a fallow period into the farming cycle.

The Cang-apa family history provides further evidence of the role of population growth in contributing to land-use pressures. Those living members from several generations choosing to remain in Siquijor's rural areas appear to contribute significantly to enlarging the extended family

---

7 Despite the decreased American presence, market demand for cattle remains strong, as indicated in continued net imports into the Philippines, a situation attributed to the country's growing, affluent middle class.

8 See G. Hardin, "The Tragedy of the Commons," *Science* 162 (1968): 1243-1248.

size, many of whom, one assumes, remain dependent on the land. This in turn contributes to increased pressure, compounding the problems of an estimated average nuclear family size of six.

Emigration, despite its significant role both in reducing Siquijor's rate of population growth and as a source of remittance income, appears to have failed to arrest the decline in farm size resulting from land fragmentation.

Despite the larger than expected percentage of absolute tenancy, there is little evidence of changes in trends over the last fifteen years. Moreover, there do not appear to be substantial differences in land-use practices between tenants and owner-operators, as indicated by the high degree of erosion severity common to most of the farms sampled. Similarly, trends in parcelization or conversely consolidation of land do not appear to be factors in upland land-use pressures. This view contrasts with the San Carlos study, which indicated a substantially higher degree of parcelization in Siquijor's highland farms. While this may be explained by use of official figures in the Cang-apa land fragmentation research, the questionnaire data indicate that only 32 out of 102 farms were divided into parcels, which provides further evidence supporting the present study's tentative conclusion.

National, provincial, and even local policies appeared of little consequence in the conflict. However, one could argue that the Pasture Improvement Program, a program created to increase livestock productivity and improve pastures, if successfully implemented in Siquijor, not only could accomplish its stated objectives but, if regarded by the upland community farmers as a valuable land improvement, could be used to introduce controlled-grazing strategies. The failure of the program to have any discernible impact in Siquijor underscores its contribution to the problem. Perhaps of even greater interest was the widespread public disinterest in local legislation enacted to provide relief for illegal grazing on private lands.

Discussion

Theoretical Considerations

Two hypotheses are given common currency in explaining land degradation in many developing countries. The first interprets it as a direct result of ignorance among local land users. This explanation seems to have fallen into disregard as more sophisticated research has demonstrated the contribution of such factors as cultural ethos, perception, and economic circumstances to the problem.[9] The second hypothesis attributes degradation to population pressure and is based on many of the Malthusian arguments

---

9 See Piers Blaikie and Harold Brookfield, "Approaches to the Study of Land Degradation," in *Land Degradation and Society*, ed. Piers Blaikie and Harold Brookfield (London: Methuen and Co., 1987), pp. 34-36

most recently attributed to the *Limits to Growth* "pessimists." The hypothesis has been criticized, among other things, for its failure to explain erosion in underpopulated areas like Australia and for its contradictory evidence of lands responding positively to additional labor inputs.[10]

In addition to these, a third hypothesis attempts to explain soil degradation in terms of surplus extraction. It argues that market-dictated surpluses are extracted from cultivators (or cattle owners in the case study) who over time have become marginalized and in desperation are increasingly forced to extract surpluses from the environment. In this case marginalization is defined as the farmers' loss of the ability to control their own lives through their forced incorporation into the world economic system.[11]

The study appears to support in varying degrees all three explanations. Perhaps the least satisfactory of the three is the role of ignorance. The best examples supporting this interpretation are: the widespread belief that *kaingin* agriculture is contributing to soil erosion; the failure to recognize the significance of overgrazing; and ignorance with regard to the nature of coral reefs and the effects of sedimentation. While one could argue that the study identified lack of awareness as among the human factors contributing to erosion and as a potential impediment to the intrabasin consensus needed for issue resolution, it hardly appears to be a fundamental factor in explaining the general situation.

A much stronger argument appears to exist for the role of population in contributing to the island's degradation, as evident in the results derived from the fragmentation and family history studies. Despite emigration, there can be little doubt that increasing population has contributed to the declining average size of the farm and increased the average size of the family dependent on same.

However, the third hypothesis appears to go farthest in explaining the case study data. A Marxist resource economist might explain the present situation in Siquijor as the result of cumulative influences of a capitalist market economy exerted through time beginning with its initial imposition some four hundred years ago. A national policy promoting participation in international trade complemented by programs designed to encourage local production of indigenous and exotic crops suitable for export in turn served as the means to integrate local livelihood systems into the national and, implicitly, international market economy. Shifts in these markets left local producers with few alternatives except to follow suit. More recently, declines in traditional export markets in the face of growing population and

---

10 Ibid., pp. 27-34.

11 See Blaikie, *Political Economy*, p. 125.

its attendant effects, manifested in land fragmentation and loss of insular self-sufficiency in staples, have marginalized the farmer.

This has forced farmers to intensify land use, join the labor market, and/or emigrate either to continue farming in less stressed environments or join off-island labor-markets. The situation whereby a gap between money required to purchase staples and income received from the selling of market commodities causes intensification of commodity production has been termed the "reproduction squeeze."[12] Moreover, the process has become institutionalized as a significant number of the farmers' children will stay in farming owing to an absence of options. One result of these market-dictated forces was the establishment and subsequent growth of Siquijor's cattle industry. Surpluses accumulating from the cattle industry, though undocumented in the present study, are in all likelihood accruing to middlemen (buyers, ship owners and operators, feedlot owners, merchants of agrochemicals, managers of abattoirs, and the retail industry) rather than the farmer, resulting in an intensified production process.

Moreover, the forced departure of individuals from farm lands is playing an exacerbating rather than a mitigating role in degradation through its reduction of the insular labor pool, and possibly through the influence of offshore remittances. Both factors contribute to a gradual but chronic decline in traditional land-use technologies, artificially prolonging their abandonment and/or delaying the adoption of more radical transformations. Boserup's concept of "induced innovation" suggests that intense land pressure can provide the required stimulus to promote the development of local technologies that mitigate the effect of population pressure on the land through increased efficiencies in production. The lessening of this pressure through emigration may, one can argue, reduce both the urgency and rate of technology development. While the case study provides some evidence of indigenously developed land conservation technology in the form of the omnipresent rock terrace, it was otherwise rare and certainly has not been sufficient to offset the problem. However it must be noted that this doesn't necessarily support the application of Boserup's concept to Siquijor. It has been observed that rates of change among factors affecting demands on soil can outstrip the indigenous population's ability to induce technological change.[13]

Finally, with respect to economic development, the government, for any number of reasons, has favored other provinces and regions in allocation decisions, to Siquijor's detriment. There are exceptions to this general-

---

[12] See Henry Bernstein, "African Peasantries: A Theoretical Framework," *Journal of Peasant Studies* 6 (July 1979): 427.

[13] See Ester Boserup, *The Conditions of Agricultural Growth* (Chicago: Aldine Publishing Co., 1965) and Blaikie, *Political Exonomy*, p. 25.

ization, of which the CVRP is perhaps the best example; however, there is a general perception in the Philippines, particularly in the provinces, that communities at the center (Manila/Luzon) benefit at the expense of those at the periphery, particularly in terms of provision of government services and infrastructure. This can be attributable to a number of causes, including culture, physiographic constraints, and corruption.

This hypothetical Marxist interpretation attributing the responsibility for degradation to excesses in the market economy does not adequately account for several nonmarket factors that appear to play a role in Siquijor, including population, social guidelines (most prominent in local tolerance of open grazing), and physical constraints. These factors do not invalidate the Marxist explanation, but provide further evidence of the complexity of the issue.

The study's results indicate that the model proposed in the research design to explain the situation in Siquijor was deficient in a number of respects. First and perhaps most important, it failed to account for the central role of external market forces, both historically and in the more recent past, in contributing to present land-use practices. Similarly, the role of local customary practices was not considered. Although overpopulation could not be confirmed as a factor contributing to encroachment on marginal lands, it did contribute to increased pressure on the land and thus to poor land-use practices. Despite the existence of a common and generally accurate perception among upland and coastal communities of the erosion/sedimentation issue, no correlations could be drawn between the perceptions and adoption of adjustments among coastal resource user groups. Finally, there was some evidence that selected government policies played a role, albeit one of neglect, in the issue.

In a broader context, it is interesting to compare the Siquijor case study to a model developed to explain soil erosion in marginal lands of former colonies. The model has five stages that are linked to each other and mutually reinforcing. These are a reduction of traditional trade among hill people, creating pressure on agriculture to provide a source of income; out-migration of much of the population to serve as a surplus labor reservoir; subsequent receipt of remittances, serving to prop up the economy; absence of differentiation of traditional agriculture characteristics in contrast to more productive areas directly under colonial rule; and increased population pressure owing to declines in death rates, resulting in a diminution and often fragmentation of holdings, in turn increasing pressure on the surrounding resources. These factors in combination contribute to the loss of the inhabitants' self-sufficiency.[14] Although the present case study has

---

14 See Blaikie, *Political Economy*, pp. 4, 35, 135-137, and 240.

not developed all the information required to test the model, there is some obvious evidence for its general relevance to Siquijor.

Implications for Management

The Siquijor case study clearly demonstrates the vulnerability of coastal resource user systems to a river basin's physical and social processes. Many of the more commonly perceived management tools useful for mitigation of coastal resource conflicts (e.g., buffer zones, setbacks, zoning, and permits) appear inadequate to address the problems of offsite sources. These institutional responses, designed primarily to mitigate sediment-related issues at the site of impact, when applied to conflicts with upland sources are often reactive in nature and fail to address the source of the problem. A number of management approaches possibly capable of addressing this type of conflict, however, do exist. Two of these, broad-scope sectoral planning and regional planning, could be used to modify incompatible land- and water-use practices that in combination result in conflict.[15] Interest in harmonizing land and water uses within a watershed framework is a concept that dates back to early nineteenth-century Europe.[16] The concept of articulated watershed management is based on the explicit recognition of the interdependency between ground cover, land use, and hydrology, and argues that any attempt at river basin management must take into account both land and water use in an integrative framework.[17] Despite its narrow acceptance in the United States, the concept has been widely used in tropical developing countries.[18] Moreover, since White's writings, evidence appears

---

[15] Jens C. Sorenson, Scott T. McCreary, and Marc J. Hershman, *Coasts: Institutional Arrangements for Management of Coastal Resources* (Columbia: Research Planning Institute, 1984), pp. 41-80.

[16] While it is unclear when the concept was first formulated, Gilbert White suggests that it first occurred in western Europe at or near the beginning of the nineteenth century. See Gilbert F. White, "Comparative Analysis of Complex River Development," in *Environmental Effects of Complex River Development*, ed. Gilbert F. White (Boulder: Westview Press, 1977), pp. 15-16.

[17] Interest in its use in the United States grew belatedly partially as a result of less severe erosion problems. The concept's principal applications have been in coordinating water and related land uses for agricultural production and pollution abatement and flood prevention in urban areas. With these exceptions it has rarely been put into action. Explanations cited for its slow acceptance include an absence of scientific facts demonstrating conclusively the existing linkages between land use and water flow (as of 1957) and the failure to demonstrate that land-use remedies are superior to engineering works. See ibid., and U.N., Economic Commission for Europe, *Seminar on River Basin Management* (New York: United Nations, 1971), pp. 20-21.

[18] Typically development projects would support reforestation schemes in denuded lands as a means of protecting critical watersheds and project-related downstream uses and infrastructure, most notably hydroelectric reservoirs (e.g., the Cauca and Gal Oya in Colombia and Sri Lanka, respectively). See Ludwik A. Teclaff, *The River Basin in History and Law* (The

to be accumulating that supports the concept's basic premise. For example, it is argued that land use is largely controlled by water availability and that in turn land-use practices affect the partitioning of the water input through changing water losses attributed to interception and evapotranspiration and in turn water yield. Similarly, changes in land use affect soil storage capacity and in turn appear to affect seasonality of the flow and stage heights. There also appears to be increasing evidence supporting certain interdependencies between reforestation schemes and downstream hydrological characteristics.[19] As a result of the growing, albeit still limited, understanding of the close interrelationships between land and water use new efforts are being made to reexamine the usefulness of the articulated approach in watershed management.[20]

Some technical means exist to bring about compatible land and water uses. Examples in upland areas include: the use of controlled grazing (increased use of corrals, cut-and-carry feeding techniques, and use of fences); pasture improvement (introduction of nutritious grasses, controlled use of burning); mitigation of dry season constraints (increased use of silage and feed supplements); increased control of physical processes contributing to erosion (use of drainage canals, hedgerows, rock terraces); and the restoration of soil fertility and vegetative cover in the landscape (use of legumes, inter- and cover cropping). Downstream technologies conceivably would include a wide range of hard flood-control works (e.g., canalization, impoundments, dikes); soft solutions (e.g., crop loss insurance); and novel approaches providing for a broader range of adjustment choices available to the coastal community (e.g., creation of new productive habitat in the form of artificial reefs, promotion of alternative livelihoods).

Moreover, the review of institutional arrangements indicates that the Philippines is favored by a wealth of legal options that could facilitate the implementation of these and other technologies through well-designed planning.

However, the case study has also demonstrated that such approaches need to be broadened to accommodate differences in community perceptions of the problem; differences that must be addressed if consensus is to be reached for the development of a basinwide approach to issue resolution. It

---

Hague: Martinus Nijhoff, 1967), p. 2; and C.H. MacFadden, "The Gal Oya Valley: Ceylon's Little T.V.A.," *Geographical Review* 44 (April 1954): 271-281.

19 See Malin Falkenmark, "Integrated View of Land and Water," *Geografiska Annaler* 63 A (July 1981): 265; and Lawrence S. Hamilton, *Strategies, Approaches and Systems in Integrated Watershed Management* (Rome: United Nations Food and Agricultural Organization, 1986), pp. 33-51.

20 Falkenmark, "Integrated View," p. 267.

has also demonstrated how local beliefs and traditions (i.e., fatalism and open grazing) can affect land-use practices.

Perhaps of even greater importance is the need to avoid the pitfalls associated with the misuse of the aforementioned management tools. They are reactive in nature in failing to address the underlying sources of the problem, which in the last analysis concerns questions of social equity.

These characteristics of the problem argue for institutional flexibility that can be achieved only through community-based approaches. However, given the significance, magnitude, and complexity of the underlying driving forces contributing to environmental degradation, such approaches would appear to be inadequate in the absence of outside public sector support. To facilitate obtaining this support three prerequisites have been identified: the need for a shared perception of the problem between land users and management institutions to ensure an equitable sharing of costs and benefits associated with conservation; a means to keep the producers and officials in close contact, the latter supposedly with the ability to prescribe the appropriate answers to the problems; and a collective discipline in implementing the conservation work.[21] Small demonstration projects such as the CVRP pilot activities are one means, albeit time consuming, to promote a community/institutional environment capable of meeting these criteria. Key elements associated with such approaches include policy reform, education, training, public participation, skillful use of incentive programs, and the judicious application of appropriate technologies.

Conclusion

The case study has demonstrated that negative environmental impacts associated with poor upland land-use practices in combination with a watershed's biophysical processes can affect coastal land and water uses. The significance of these findings lies in the degree to which they are representative of a larger problem. In the author's view, the situation in Siquijor reflects a situation common in other tropical locales where well-defined watersheds with highly productive coastal habitat and/or coastal land uses are characterized by rural economies dependent on intensified exploitation of the land. The gravity of the situation rests in the scale of degradation of marginal lands and the difficulties in finding solutions where social equity issues are the underlying cause of the problem.

In light of the dearth of information and understanding concerning these intrabasin conflicts, it is hoped that the case study has provided a basis from which to design subsequent research. The integrative research framework identifying ecological systems and populations is promising. The

[21] Ibid., pp. 154-155.

common use of a watershed as an ecosystem however needs to be expanded to include the nearshore environment. Studies directed to the selection of issue-specific criteria and boundary definition of this environment would be useful. Similarly, additional work is needed in refining the definition of the coastal population. In the case of Siquijor, the population was dependent on multiple livelihoods and comprised one community rather than individual groups dependent on different sources of incomes. This has obvious relevance to perception studies and also needs to be addressed in the use of the human-use ecosystem paradigm; i.e., if a population is not dependent on one ecosystem as a source of livelihood, the identification of the appropriate regional population may be required for the study. Additional frameworks to meet specific research needs should be developed and tested.

In regard to Siquijor specifically, and the Central Visayas and other similar situations generally, a number of critical data gaps should be addressed to complement existing information. One such gap concerns the role of remittances in contributing to land degradation. Specifically, to what degree is the existing population being supported by remittances: is the insular population living beyond the margin as a result of external sources of income? Is offshore income being invested in cattle as an alternative to the bank, and if so, is it another driving force contributing to overgrazing? Another question relates to the pattern of receipt of remittances over time and its correlation to changes in land-use practices and attitudes toward conservation. Finally, what have been the ensuing adjustments to the cessation of remittances among the members of the insular population and what is the significance to the general welfare if this occurred on a large scale (as in the case of a recession or return of guest workers)?

A second critical need is to better define the role of the middlemen and their relationship to surplus extraction. These will differ from sector to sector, but they represent an important element to the Marxist explanation for environmental degradation among the rural poor that was not addressed in the present study.

With regard to the research itself, the data and analysis could be improved upon significantly through additional studies. Comparative studies could be designed to assess: the importance of watershed size in shaping intrabasin community perceptions, particularly with regard to downstream community impacts; the similarities and differences in patterns of rural-based environmental degradation associated with regional economies driven by different national or international markets; and the downstream/upstream impacts, social dynamics, perceptions, and institutional aspects associated with coastal communities migrating up the basin to the detriment of traditional upland communities. Where relevant, said studies should be implemented over a period of time sufficient to judge how perceptions vary with the recency and severity of hazard. One wonders how the results would differ, particularly in the coastal community perception studies, if the Siquijor study year had been characterized by torrential rains and flooding rather than drought.

Parallel to this should be studies to determine if spatial patterns can be determined from the responses of coastal users and property at risk. For example, do coconut growers and paddy farmers subject to greater river bank erosion and flood risk respectively demonstrate increased understanding of the processes and adjustment characteristics of the hazard than those less at risk?

In the subsequent approaches to the topic, interested researchers should endeavor to obtain a sufficient budget to develop a complete data base. Improvements on the existing study's equipment might include a continuous recording stream gauge, an integrated suspended-sediment sampler, and adequate equipment to complete a survey of currents. Care should be taken to avoid an overly complicated and long survey questionnaire.

Finally, with regard to Siquijor, one cannot help but conclude that the situation is both serious and likely to worsen. The rural society appears to have been forced into using its land (and water) resources beyond most accepted definitions of sustainability. Critical pressures and constraints, particularly population growth, land fragmentation, and the absence of options for a significant percentage of the population, indicate that the problem will grow. Moreover, it is a population highly at risk from its apparent dependence on outside income and limited external opportunities for employment. Despite the existence of technologies that can promote intensification of land use compatible with conservation and despite some evidence for their successful application in the island, it is difficult to be optimistic in the absence of major reforms in Filipino government attitudes, policy, and approaches to the rural poor.

Appendix 1: Upland Community Questionnaire

Below are a series of questions asking the respondents to describe their feelings regarding their families' general well-being, current farming and fishing trends and practices, environmental perceptions, certain government policies, and recent climatic patterns. The questions were multiple choice, yes/no, response lists, or open-ended in nature.

1. Are you or have you ever participated in either the CVRP or DA/CEDP Projects?
2. From the time you began farming your land in Maria would you say your general economic situation has improved for you and your family?
3. From the time you began farming your land in Maria would you say your present farm conditions have improved?

Since you began farming your lands in Maria would you say your:

4. Pasturelands are improving?
5. Pasturelands are more plentiful?
6. Corn production is increasing?
7. Livestock production is getting better?
8. Livestock are getting fatter?
9. Livestock are more healthy?
10. Livestock are more plentiful?

Since you began farming your lands in Maria would you say:

11. Loss of your soils from rain and runoff is getting worse?
12. Your soils are becoming less productive?
13. Flooding of your lands is getting worse?

The government has initiated several programs designed to help small rural farmers (e.g., the Livestock Dispersal Program, Pasture Improvement Program). Would you say the government has helped you in:

14. Building your cattle herd?

Note: The presentation of the questionnaire has been slightly altered for purposes of economy of space. The content and sequence of questions however remain unaltered.

15. Improving your pasture?
16. Building your soil fertility and stopping soil erosion?

From your past cumulative experience living in Siquijor would you say that for the calendar year 1987:

17. The rainy season began early?
18. The rainy season began late?
19. The dry season came early?
20. The dry season came late?
21. The rains were more frequent?
22. The rains were less frequent?
23. The rains were more intense?
24. The rains were less intense?
25. Do you wish to stay in farming?
26. Do you hope your children will farm your lands when they grow up?
27. Is your farm divided into parcels?
28. Do you own any of these?
29. Are you a tenant on any of these?
30. If you don't presently own the land you farm do you rent or lease or have an alternative arrangement with the owner(s) in exchange for the use of this land?
31. If you don't presently own the land you farm did you ever own your own farm?
32. Do you own your cattle in their entirety?
33. Do you own only a share or a portion of your cattle on the farm?
34. Are their other areas in the Maria River basin available to graze your cattle?
35. Do you own other agricultural lands in the Maria River basin?
36. Do you fish in Maria Bay?
37. Have you participated in the Department of Agriculture Bureau of Animal Industry's Livestock Dispersal Program?
38. Have you participated in the Department of Agriculture Bureau of Animal Industry's Pasture Improvement Program?
39. Which of the following conservation measures do you use on your lands?

Livestock	Agriculture production	
Cut and carry	Gully controls	Strip cropping
Controlled grazing	Intercropping	Careful or no tillage
Feed supplements	Crop rotation	Mulching
Crop debris as feed	Cover cropping	Other
Silage	Contouring	
Improved pastures	Fertilizers	
Controlled breeding	Terracing	
Other	Hedgerows	

40. From the time you began farming your land in Maria has your total farm size increased/decreased/or stayed the same?
41. From the time you began farming your land in Maria has your total pastureland increased/decreased/or stayed the same?

42. From the time you began farming your land in Maria has your total cattle herd increased/decreased/or stayed the same?
43. Have you sold a portion of your land in the last 5/10/15 years?
44. Have you bought a portion of your land in the last 5/10/15 years?
45. How often do you put your pasturelands in fallow? (once every 1/2/3/>3 years)
46. How much land do/did you own in the present/last 5/10/15 years?
47. How much pasture do/did you own in the present/last 5/10/15 years?
48. How many cattle do/did you own in the present/last 5/10/15 years?
49. Check the type and number of livestock under appropriate feeding practices (for cattle, carabao, and goats).

    Tethering
    Pasturing
    Cut and carry
    Hay fed
    Silage
    Concentrate
    Others

50. If you are presently using lands owned by an absentee owner, how often does he return to oversee his lands?
51. Kindly estimate your total existing land area by tenure for:

    Pasture lease (government land)
    Lease/rented from private owner
    Temporarily occupied public land
    Personally owned titled land
    Tenanted land
    Others (state)
    Total

52. Kindly estimate your present total land area by uses for:

    Irrigated riceland
    Rain-fed cornland
    Rain-fed crop mix
    Pastureland
    Tree farm
    Others (state)
    Total

53. How much land have you sold/purchased since you acquired your land for :

    Irrigated riceland
    Rain-fed cornland
    Rain-fed crop mix
    Pastureland
    Tree farm
    Others (state)
    Total

54. Kindly estimate your total land area by gradient for:

    Steep/stony/rocky
    Steep but cultivable
    Rolling and cultivable
    Rain-fed flatland
    Irrigated flatland
    Irrigated lowland
    Others (state)
    Total

55. Kindly estimate approximate distances between your household and the closest:

    All-weather road
    Dirt road
    Barrio proper
    Town proper
    Public market
    Perennially running river

56. What most nearly describes frequency of public utility vehicle or other transportation serving your area?
    - Very frequent and regular
    - Frequent but not regular
    - Infrequent and irregular
    - None at all
57. Why do you say your general farm conditions are improving/declining?
58. Why do you/do you not want your children to continue farming?
59. What type of grass (grasses) do you graze your cattle on?
60. How often do you sell your cattle, how old are they, and for what purpose?
61. Which do your prefer for farming, cattle or carabao, and why?
62. If you don't own your pastureland in its entirety, what is the nature of the arrangement whereby you can graze your cattle on the owners' land?
63. How many crops of corn do you grow each year?
64. If you own your land, how did you come by way of ownership?
65. If you lease or are otherwise not the owner of the land you farm, please describe the nature of the arrangements with the owner(s) in exchange for the use of this land.
66. If you owned land previously, why are you now renting/leasing?
67. Why have you reduced/added to your original landholdings since you first started farming in Maria?
68. Why have you reduced/added to your pasturelands since you first started farming in Maria?
69. Why have you reduced/added to your cattle herd since you first started farming in Maria?

69a. Why don't you grow fruit trees on your pasturelands?

70. Your three most severe farm problems in order of priority are:
71. The reason(s) you have poor soils is/are:
72. You try to stop your soil loss by:
73. What you would like to do to stop soil loss is:
74. You can't do these because:
75. How many stone wall terraces do you have on your lands?
76. When were these built and by whom?
77. Why do you think you are losing your soil?
78. Where does your soil go?
79. How does it get there?
80. Who is affected/utilizes your soil?
81. What do they do with it?
82. Whose responsibility is it to prevent loss of these soils?
83. How do you adjust for a food shortage in feeding your cattle?
84. Describe the three worst cases of loss of soils and crops in Maria (please provide dates, description of the event, and damages).
85. The government could help you prevent the loss of your soils by:
86. How has the government assisted you in improving your livestock?

87. How has the government assisted you in improving your pasture?
88. How has the government assisted you in reducing your soil loss and fertility decline?
89. Please explain briefly any problems you see in the Livestock Dispersal Program implemented through the Department of Agriculture's Bureau of Animal Industries.
90. How many cattle would you estimate you have received through this program?
91. Please explain any problems you see in the Pasture Improvement Program implemented through the Department of Agriculture's Bureau of Animal Industries.
92. How old are you?
93. When did you begin farming the land you presently work?
94. How many children do you have?
95. What will the mode of inheritance be in passing your farm down to your children? How did you decide who gets what? What arrangements exist or will you make for cattle inheritance?
96. How did you inherit your lands? (How many brothers/sisters were there; was land distributed by equal shares; if not, how; are the inherited parcels in the heirs' or parents' [or other] names?)
97. If you own other farms in the Maria watershed, where are they (*barangay/sitio*) and what do you grow/graze?
98. If you own only a portion of your lands in the Maria, who are the other owners (name/relation)?
99. If you are presently farming the land of an absentee owner, may we contact him for purposes of the survey?
100. If so, what is his/her name and how can we reach them?
101. May we return to ask you additional questions if needed?

Appendix 2: Coastal Community Questionnaire

Below are a series of questions asking the respondents to describe their feelings regarding their family's general well-being, current farming and fishing practices, environmental perceptions, certain government policies, and recent climatic patterns. The questions were multiple choice, yes/no, response lists or open-ended in nature.

1. Are you or have you ever participated in either the CVRP or DA/CEDP Projects?
2. Do you grow paddy in the Maria watershed?
3. Do you own/harvest coconut trees in the Maria watershed?
4. Do you fish in Maria Bay?
5. Do you own farmland elsewhere in the upland areas of the Maria watershed?
6. Do you have other means to support you and your family?
7. What is your principle source of livelihood from the list above?

From your past cumulative experience living in Siquijor would you say that for the calendar year 1987:

8. The rainy season began early?
9. The rainy season began late?
10 The dry season came early?
11. The dry season came late?
12. The rains were more frequent?
13. The rains were less frequent?
14. The rains were more intense?
15. The rains were less intense?
16. From the time you began growing paddy in Maria would you say your general economic situation has improved for you and your family?

Since you first began growing paddy in Maria would you say your:

17. Paddy production is increasing?

Note: The presentation of the questionnaire has been slightly altered for purposes of economy of space. The content and sequence of questions however remain unaltered.

18. Water supply is plentiful?
19. Fertilizer is adequate?

Since you first began growing paddy in Maria would you say your paddy fields are increasingly being affected by :

20 Pests?
21. Drought?
22. Flooding?
23. Siltation?

Has the government helped you:

24. Increase your paddy production?
25. Improve your water supply?
26. Meet the problem of flooding of your paddy?
27. Meet the problem of siltation of your paddy?

28. Are your paddy fields divided into parcels?
29. Do you own any of these?
30 Are you a tenant on any of these?
31. After a flood is the paddy covered with large amounts of silt?
32. Does this affect your paddy production?
33. From the time you began growing paddy have you had to increase the frequency of cleaning of your water diversion canals?
34. From the time you began growing paddy have you had to build defense works to protect your fields from flooding?
35. When did you first begin growing paddy in the Maria watershed?
36. Do you produce enough paddy to sell or do you use it all for personal consumption?
37. Why is your paddy production increasing/decreasing?
38. If you don't own the paddy you are growing in its entirety, please describe the nature of the arrangement under which you work.
39. What is/were the rice varieties you use/have used in the present/last 5/10/15 years?
40 Why are you using the present variety?
41. What are the three most serious concerns facing you as a paddy farmer, listed in order of severity?
42. How many times in a year does a storm result in flooding and sediment covering your rice?
43. Where does the sediment come from?
44. What is the cause of sedimentation?
45. How does it get to your fields?
46. Does sediment help your fields (if so, how; if not, why not)?
47. Why are other people losing their soils?
48. Who is responsible for the silting of your fields?
49. Whose responsibility is it to solve the problem of sedimentation and flooding?
50 What measures exist to prevent the flooding and the siltation of your paddy?
51. Which ones do you use?
52. Why not use the other measures you list?

53. What were the three worst storm events occurring in your memory resulting in flooding and siltation on your lands? (Please describe damages, give dates, and characterize.)
54. Please identify the approximate location of your fields on the accompanying map. (Also kindly provide *barangay* and *sitio*.)
55. What has the government done to assist you in reducing the problem of flooding? Of siltation?
56. From the time you began harvesting coconuts in Maria would you say your general economic situation has improved for you and your family?
57. Since you first began harvesting coconuts in Maria would you say your trees are becoming more productive?

Since you first began harvesting coconuts in Maria would you say your trees are increasingly being affected by:

58. Pests?
59. Seasonal storms?
60. Erosion and the caving-in of river banks?
61. Has the government helped you increase coconut production?
62. Has the government helped you with pests?
63. Has the government helped you stop river bank erosion affecting your trees?
64. Do you own the coconut trees you harvest in their entirety?
65. If not, do you own some portion of them?
66. Do you own the lands on which the coconut trees grow?
67. Do you harvest coconut trees located on the river's edge?
68. Has the loss of trees adjacent to river banks been getting worse since you first began to harvest coconut?
69. When did you first start harvesting coconuts in Maria?
70 If the trees and accompanying lands are not yours in their entirety, please describe the arrangements under which you harvest the coconuts.
71. Why is your coconut production decreasing/increasing?
72. What are the three most serious concerns facing you as a coconut producer, listed in order of their severity?
73. Why are the river banks eroding and affecting coconut trees?
74. What is the cause of eroding river banks?
75. Where does the sediment come from?
76. What is the cause of the sedimentation?
77. How does it get to the river banks?
78. Why are other people losing their soils?
79. Who is responsible for the bank erosion?
80 Whose responsibility is it to solve the problem?
81. Does this affect your personal coconut harvest?
82. What measures exist to prevent the loss of coconut trees adjacent to these eroding river banks?
83. If applicable, which ones do you use?
84. If applicable, why not use the other measures you list?

85. What were the three worse storm events resulting in losses to your coconut harvest? Please describe (damages, dates, characterization).

86. Please identify the approximate location of your trees on the accompanying map. (Also kindly provide *barangay* and *sitio.*)

87. Please describe what the government has done to assist the coconut farmers in reducing loss of coconut trees associated with bank erosion.

88. From the time you began fishing in Siquijor would you say your general economic condition has improved for you and your family?

Since you first began fishing in Maria Bay would you say your:

89. Fish catch is increasing?

90 Fish are plentiful?

Since you first began fishing in Maria would you say:

91. Your daily fish catch is becoming less?

If your fish catch is declining do you believe it is due to:

92. Too many fishermen?

93. The use of explosives and poisonous chemicals killing the coral reefs?

94. Increasingly river-borne sedimentation killing the reefs?

95. Other?

96. Has the government helped you to increase fish production?

97. Has the government helped you in meeting the problems of declining fish catch?

98. Do you fish elsewhere besides in Maria Bay?

99. Do you have a motorized boat?

100. Do you fish predominately at night?

101. Do you fish predominately during the day?

102. Do you consider yourself to be a full-time fisherman?

103. From the time you first began fishing have you had to increase the total daily time spent to catch the same amount of fish?

104. From the time you began fishing have you had to go farther to catch the same amount of fish?

105. From the time you first began fishing have you had to switch to different fishing gear to catch other species of fish owing to declining yields of your preferred species?

106. After a storm is the water dirty with sediment?

107. Does this affect your fish production?

108. Are corals living animals?

109. Does your livelihood as a fisherman depend on healthy coral reefs?

110 When did you first begin fishing in Maria Bay?

111. Do you produce enough fish to sell or do you use it all for personal consumption?

112. Why is your fish catch increasing/decreasing?

113. Please describe your preferred fish species for daily and seasonal conditions, equipment used, and predominate zone fished (see map).

114. Please identify the approximate location of your most usual fishing grounds over time (refer to the accompanying map).

115. If you have shifted fishing zones over time, why have you?
116. What is the predominate type of gear you would use in the following monsoon/zone combinations? (Refer to accompanying map.)
117. Why don't you fish those zones left vacant in #116 (if applicable)?
118. When and why do you go outside of Maria Bay waters to fish?
119. What are the prevailing currents in Maria Bay during the following tidal/monsoonal conditions?
120 What are the three most serious concerns facing your future as a fisherman? Please list in order of severity.
121. On an annual basis how often following a storm does the Maria Bay get dirty with sediment?
122. Where does the sediment come from?
123. What is the cause of sedimentation?
124. How does it reach the fishing waters?
125. Does sediment help your fish catch (if so, how; if not, why not)?
126. How does it affect coral reefs?
127. Why are other people losing their soils?
128. Who is responsible for the sediment?
129. What should be done about it as a problem?
130 What measures exist to avoid the effects of dirtying Maria Bay waters?
131. Which do you use?
132. Why not use the other measures you list?
133. What were the three worst storm events resulting in siltation in Maria Bay (please describe damages, dates, and general character)?
134. Where does the sediment most affect the bay following a storm?
135. How has the government helped you in reducing your fish losses?
136. How has the government helped you in reducing the problem of sedimentation?
137. How old are you?
138. May we return to ask you additional questions if needed?

Appendix 3: Families and Species Identified in Fish Census of Maria Bay

Acanthuridae
- *Acanthurus bleekeri*
- *Ctenochaetus striatus*
- *Naso minor*
- *Zebrasoma scopas*

Apogonidae
- *Apogonsp.*
- *Cheilodipterus macrodon*
- *C. quinzuelineatus*

Aulostomidae
- *Aulostomus chinensis*

Caesionidae
- *Pterocaesio pisang*

Centriscidae
- *Aeoliscus strigatus*

Chaetodontidae
- *Chaetodon aurega*
- *C. kleini*
- *C. vagabundus*
- *Heniochus varius*

Dasyatidae
- *Dasyatis kuhlii*

Haemulidae
- *Plectorhynchus chaetodonloides*
- *P. pictus*

Lutjanidae
- *Lutjanus decussatus*
- *L. fulvus*
- *L. lutjanus*
- *L. sebae*

Gobiidae

Labridae
- *Bodianus mesothorax*
- *Cheilinus celebicus*
- *C. undulatus*
- *C. bimaculatus*
- *C. fasciatus*
- *C. trilobatus*
- *Cirrhilabrus sp.*
- *Gomphosus varius*
- *Halichoeres sp.*
- *H. centiquadrua*
- *H. hoeveni*
- *H. poecilopterus*
- *H. prosopeion*
- *Labroides dimidiatus*
- *Stethojules sp.*
- *Stethojules trilineata*
- *Thalassoma hardwickei*
- *T. lunare*

Mullidae
- *Parupeneus barberinus*
- *P. trifasciatus*

Mugilsididae
- *Parapercis sp.*

Nemipteridae
- *Scolopsis bilineatus*
- *S. ciliatus*

Paralichthyidae

Plotosidae
- *Plotosus lineatus*

Pomacanthidae
- *Centropyge vroliki*
- *C. tibicen*
- *C. bicolor*
- *Chaetodotoplus mesoleucus*
- *Pygoplites diacanthus*

Pomacentridae
- *Amblyglyphidodon curacao*
- *Amphiprion clarkii*
- *A. frenatus*
- *A. perideraion*
- *Anthias mortoni*
- *A. tuka*
- *Chromis analis*
- *C. elerae*
- *C. flavormaculatus*
- *C. ternatensis*
- *Dascyllus aruanus*
- *D. melanurus*
- *D. reticulatus*
- *D. trimaculatus*
- *Dischistodus fasciatus*
- *Glyphidodontops cyaneus*
- *G. hemicyaneus*
- *G. leucopomus*
- *G. rollandi*
- *Paraglyphidodon melas*
- *P. nigrosis*
- *Plectroglyphidodon lacrymatus*
- *Pomacentrus alexanderae*
- *P. amboinensis*
- *P. bankanensis*
- *P. coelestis*
- *P. moluccensis*
- *P. philippinus*
- *P. vauli*

Scaridae
- *Cetuscarus bicolor*
- *Scarus sp.*

S. fasciatus
S. scaber

Serranidae
Cephalopholis argus

Siganidae

Synodontidae
Saurida gracilis
Synodus variegatus

Teraponidae
Terapon jarbua

Tetraodontidae
Canthigaster valentini
C. solandre
Tetrodon nigropuntatus

Zanclidae
Zanclus cornutus

Bibliography

Achay, Antonio D. "A Management Survey of Twenty Five Rice Farms in Barangay Lo-oc, Maria Siquijor." B.A. thesis, Foundation University, 1980.

Ad Hoc Committee on Geography. *The Science of Geography*. Washington, D.C.: National Academy of Sciences, National Research Council, 1965.

Associacion ng mga Consultants na Independente (Philippines), Inc. "Central Visayas Regional Project-1: Mid-Project Review Report." Cebu City: 1986. Typewritten.

Ayson, G., and Abletez, J.P. *Barangay: Its Operations and Organizations*. Manila: National Book Store, 1987.

Bacalso, Jovencio M. "The Cattle and Carabaos in Los Banos, Calamba and Cabuyao, Laguna, after the Liberation of the Philippines in 1945." *Philippines Agriculturalist* 35 (March 1951): 163-169.

Barrows, Harlan H. "Geography as Human Ecology." *Annals of the Association of American Geographers* 13 (March 1923): 1-14.

Bernstein, Henry. "African Peasantries: A Theoretical Framework." *Journal of Peasant Studies* 6 (July 1979): 421-443.

Berry, Brian J.L. "Approaches to Regional Analysis: A Synthesis." *Annals of the Association of American Geographers* 54 (March 1964): 2-11.

Blaikie, Piers. *The Political Economy of Soil Erosion in Developing Countries*. New York: Longman Scientific and Technical, 1987.

_______. "Natural Resource Use in Developing Countries." In *A World in Crisis*, ed. R.J. Johnston and P.J. Taylor, pp. 107-126. Oxford: Basil Blackwell, 1986.

Blaikie, Piers, and Brookfield, Harold. Introduction to *Land Degradation and Society*, ed. Piers Blaikie and Harold Brookfield, pp. xvii-xxiv. London: Methuen and Co., 1987.

_______. "Approaches to the Study of Land Degradation." In *Land Degradation and Society*, ed. Piers Blaikie and Harold Brookfield, pp. 27-48. London: Methuen and Co., 1987.

Bormann, F.H., and Likens, G.E. "The Watershed-Ecosystem Concept and Studies of Nutrient Cycles." In *The Ecosystem Concept in Natural Resource Management*, ed. George M. Van Dyne, pp. 49-78. New York: Academic Press, 1969.

Boserup, Ester. *The Conditions of Agricultural Growth*. Chicago: Aldine Publishing Co., 1965.

Bourns F.S., and Worcester, D.C. "Preliminary Notes on the Birds and Mammals Collected by the Menage Scientific Expedition to the Philippine Islands." *Occasional Papers of the Minnesota Academy of Natural Sciences* 1 (1894): 1-65.

Brookfield, Harold. "On Man and Ecosystems." *International Social Science Journal* 34 (September 1982): 375-393.

Brown, Walter C., and Alcala, Angel C. "Comparison of the Herpetofaunal Species Richness on Negros and Cebu Islands, Philippines." *Silliman Journal* 33 (January 1986): 74-86.

Campbell, D.T., and Stanley, J.C. *Experimental and Quasi-Experimental Designs for Research*. Chicago: Rand McNally, 1966.

Cheetham, Russell J., and Hawkins, Edward K. *The Philippines: Priorities and Prospects for Development*. Washington, D.C.: World Bank, 1976.

Clarke, Robin, and Timberlake, Lloyd. *Stockholm Plus Ten*. London: Earthscan, 1982.

Commonwealth of the Philippines. Commission of the Census. *Census of the Philippines: 1939*. Vol. 3: *Reports by Provinces for the Census of Agriculture*. Manila: Bureau of Printing, 1940.

_______. Department of Agriculture and Commerce. Bureau of Forestry. *Annual Report of the Director of Forestry of the Philippines*. Manila: Bureau of Printing, 1940.

Cruz, C.J. "Demographic Issues in Upland Development." Paper presented at the Workshop on a National Strategy for Sustainable Development of Forestry, Fisheries and Agriculture, Manila, March 30-31, 1987. Typewritten.

Deang, Lionel; Avila, Josephine L.; and Lirasan, Froilan. "Economic Life in Seven Upland Areas of Region VII: 1981." *University of San Carlos Research Digest Series* 12 (December 1985): 1-13.

_______. "Economic Life in Seven Upland Areas of Region VII." Paper prepared for the Office of Population Studies, University of San Carlos, Cebu City, 1981. Typewritten.

Director of Agriculture. "The Animal Disease Problem." *Philippine Agricultural Review* 1 (May 1908): 186-193.

Dorita, Mary. "Filipino Immigration to Hawaii." M.A. thesis, University of Hawaii, 1954.

DuBois, Random. "Catchment Land Use and Its Implications for Coastal Resources Conservation in East Africa and the Indian Ocean." In *Ocean Yearbook 5*, ed. Elisabeth Mann Borgese and Norton Ginsburg, pp. 192-222. Chicago: University of Chicago Press, 1985.

Dunne, Thomas, and Leopold, Luna B. *Water in Environmental Planning*. New York: W.H. Freeman and Co., 1978.

Dystra, C.J. "Engineering Geological Feasibility of Rehabilitating Buhisan Reservoir." Report prepared for the University of San Carlos Water Resources Center and the Delft University of Technology Cooperating Project, Cebu City, 1977. Typewritten.

Endean, R. "Pollution of Coral Reefs." In *5th FAO/SIDA Workshop on Aquatic Pollution in Relation to the Protection of Living Resources*, pp. 343-363. Rome: United Nations Food and Agriculture Organization, 1978.

Fabre, Jean Antoine. *Essai sur la Theorie des Torrens [sic] et des Rivieres*. Paris: Chez Bidault, 1797.

Fahim, H.M. *Dams, People and Development*. New York: Pergamon Press, 1984.

Fairbridge, R.W., and Teichert, C. "The Low Isles of the Great Barrier Reef: A New Analysis." *Geographical Journal* 111 (January 1948): 67-88.

Falkenmark, Malin. "Integrated View of Land and Water." *Geografiska Annaler* 63A (July 1981): 261-271.

Falvey, J.L. "*Imperata cylindrica* and Animal Production in South-East Asia: A Review." *Tropical Grasslands* 15 (March 1981): 52-56.

Ferguson Wood, E.J., and Johannes, R.E., eds. *Tropical Marine Pollution*, Elsevier Oceanography Series 12. Amsterdam: Elsevier Scientific Publishing Co., 1975.

Galang, Zoilo M. *Encyclopedia of the Philippines*. Vols. 5 and 6, *Commerce and Industry*. Manila: Exequiel Floro, 1950.

Geertz, Clifford. *Agricultural Involution*. Berkeley: University of California of Press, 1963.

George, Carl J. "The Role of the Aswan High Dam in Changing the Fisheries of the Southeastern Mediterranean." In *The Careless Technology*, ed. M. Taghi Farver and John P. Milton, pp. 159-178. Garden City: Natural History Press, 1972.

Glacken, Clarence J. *Traces on the Rhodian Shore*. Berkeley: University of California Press, 1967.

Glover, Sir Harold. *Soil Erosion*. London: Oxford University Press, 1944.

Gomez, E.D.; Alcala, A.C.; and San Diego, A.C. "Status of Philippine Coral Reefs." In *Proceedings of the Fourth International Coral Reef Symposium*, ed. Edgardo Gomez et al., pp. 275-282. Quezon City: Marine Sciences Center, University of the Philippines, 1981.

Haggett, Peter. *Geography: A Modern Synthesis*. New York: Harper and Row, 1972.

Hamilton, Lawrence S. *Strategies, Approaches and Systems in Integrated Watershed Management*. Rome: United Nations Food and Agricultural Organization, 1986.

Hamilton, Lawrence S., and King, Peter N. "Watersheds and Rural Development Planning." *Environmentalist* 7 (1984): 80-86.

Hardin, G. "The Tragedy of the Commons." *Science* 162 (1968): 1243-1248.

Hewitt, Kenneth, and Burton, Ian. *The Hazardousness of a Place: A Regional Ecology of Damaging Events*. Toronto: University of Toronto Press, 1971.

Huntington, Ellsworth. *Civilization and Climate*. 3d ed. New Haven: Yale University Press, 1915.

Hyams, Edward. *Soils and Civilization*. New York: Harper and Row, 1958.

Independent Commission on International Development Issues. *North-South: A Programme for Survival*. Cambridge: Massachusetts Institute of Technology Press, 1980.

International Union for Conservation of Nature and Natural Resources. *World Conservation Strategy*. Gland: International Union for Conservation of Nature and Natural Resources, 1980.

Jacks, G.V., and Whyte, R.O. *The Rape of the Earth*. London: Faber and Faber, 1939.

Kates, Robert W., and Burton, Ian, eds. *Geography, Resources, and Environment*. Vol. 1, *Selected Writings of Gilbert F. White*. Chicago: University of Chicago Press, 1986.

Kirchner, W.B. "An Evaluation of Sediment Trap Methodology." *Limnology and Oceanography* 20 (July 1975): 657-660.

Kretzer, David C. "How to Build Up and Improve a Herd or Flock." *Philippine Agricultural Review* 20 (July 1928): 215-226.

Lasker, Bruno. *Filipino Immigration to Continental United States and to Hawaii*. Chicago: University of Chicago Press, 1931.

Lomotan, B.S. "Climatic Types of the Philippines." In *Philippine Recommendations for Corn, 1970-1971*. Los Banos: Philippine Council for Agricultural Research and Development, 1970.

MacFadden, C.H. "The Gal Oya Valley: Ceylon's Little T.V.A." *Geographical Review* 44 (April 1954): 271-281.

McGregor, R.C. *A Manual of Philippine Birds*. Part 2. Manila: Bureau of Science, 1909.

Manresa, Miguel; Narciso, Pepito N.; and Silva, Abel L. "Comparative Efficiency of Pasture Management Methods." *Philippine Agriculturalist* 27 (October 1938): 343-356.

Maragos, J.E. "A Study of the Ecology of Hawaiian Reef Corals." Ph.D. thesis, University of Hawaii, 1972.

Marsh, George Perkins. *From the Earth as Modified by Nature: A New Edition of Man and Nature.* 1874. Reprint. St. Clair Shores, Mich.: Scholarly Press, 1985.

Mikesell, Marvin W. "Geography as the Study of Environment: An Assessment of Some Old and New Commitments." In *Perspectives on Environment*, ed. Ian R. Manners and Marvin W. Mikesell, pp. 1-23. Washington, D.C.: Association of American Geographers, 1974.

Miller, Carl R. *Analysis of Flow-Duration Sediment-Rating Curve Method of Computing Sediment Yield.* Washington, D.C.: U.S. Bureau of Reclamation, 1951.

Mitchell, Bruce. *Geography and Resource Analysis.* London: Longman, 1979.

Monahan, Edward C., and Monahan, Elizabeth A. "Trends in Drogue Design." *Limnology and Oceanography* 18 (November 1973): 981-985.

Nesom, G.E. "The Practicability of Supplying Native Beef to the Army." *Philippine Agricultural Review* 6 (March 1911): 119-121.

Nueva Espana. *Guia Oficial de Filipinas.* Manila: Ramirez y Giraudier, 1884.

Organization of American States. *Environmental Quality and River Basin Development: A Model for Integrated Analysis and Planning.* Washington, D.C.: Organization of American States, 1978.

Osborn, Fairfield. *Our Plundered Planet.* London: Faber and Faber, 1948.

Pattison, W.D. "The Four Traditions of Geography." *Journal of Geography* 63 (May 64): 211-216.

Payne, W.J.A. "The Role of the Cattle Industry in the Philippines." *Philippines Journal of Animal Science* 3 (December 1966): 12-29.

Powell, John Wesley. *Report on the Lands of the Arid Region of the United States, with a More Detailed Account of the Lands of Utah.* Report submitted to the 45th Congress, 2nd Session, H.R. Exec. Doc. 73, 1878.

Rambo, Terry. "Human Ecology Research on Tropical Agroecosystems in Southeast Asia." *Singapore Journal of Tropical Geography* 3 (June 1982): 86-99.

Rappaport, Roy A. *Pigs for the Ancestors.* 2d ed. New Haven: Yale University Press, 1984.

Redclift, Michael. *Development and the Environmental Crisis.* London: Methuen, 1983.

Richter, D.D.; Saplaco, S.R.; and Nowak, P.F. "Watershed Management Problems in Humid Tropical Uplands." *Nature and Resources* 21 (October 1985): 11-21.

Rogers, Caroline S., and Teytaud, Robert. *Marine and Terrestrial Ecosystems of the Virgin Islands National Park and Biosphere Reserve.* St. Thomas: U.S. National Park Service, 1988.

Rosell, Dominador Z. "Siquijor Island." *Philippine Magazine* 35 (June 1938): 418-436.

Russ, Garry. "Effects of Protective Management on Coral Reef Fishes in the Central Philippines." *Proceedings of the Fifth International Coral Reef Congress* 4 (1985): 219-224.

_______. "Distribution and Abundance of Herbivorous Grazing Fishes in the Central Great Barrier Reef. I. Levels of Variability across the Entire Continental Shelf." *Marine Ecology Progress Series* 20 (November 1984): 23-34.

Saarinen, Thomas F.; Seamon, David; and Sell, James L., eds. *Environmental Perception and Behavior.* Chicago: University of Chicago Department of Geography Research Paper no. 209. Chicago: University of Chicago Department of Geography, 1984.

Sahagun, Virgilio A. *Sediment Transport Study: Mananga River.* Cebu City: University of San Carlos Water Resources Center, 1985. Unpaged.

Salter, L.A. *A Critical Review of Research in Land Economics*. Madison: University of Wisconsin Press, 1967.

Sauer, Carl O. "Foreword to Historical Geography." *Annals of the Association of American Geographers* 31 (March 1941): 1-24.

_______. "The Morphology of Landscape." *University of California Publications in Geography* 2 (1929): 37-53.

Shaler, N.S. "The Economic Aspects of Soil Erosion." *National Geographic Magazine* 7 (1896): 328-338.

Sorenson, Jens C.; McCreary, Scott T.; and Hershman, Marc J. *Coasts: Institutional Arrangements for Management of Coastal Resources*. Columbia: Research Planning Institute, 1984.

Spencer, J.E. "The Rise of Maize as a Crop Plant in the Philippines." *Journal of Historical Geography* 1 (January 1975): 1-16.

_______. *Land and People in the Philippines*. Berkeley: University of California Press, 1952.

Spencer, J.E., and Hale, G. A. "The Origin, Nature and Distribution of Agricultural Terracing." *Pacific Viewpoint* 2 (March 1961): 1-40.

Steward, Julian. *The Theory of Culture Change*. Urbana: University of Illinois Press, 1955.

Stohr, Walter B., and Fraser Taylor, D.R., eds. *Development from Above or Below?* Chichester: John Wiley and Sons, 1981.

Sundborg, Ake. "Sedimentation Problems in River Basins." *Nature and Resources* 19 (April 1983): 10-21.

Tate, D.J.M. *The Making of Modern South-East Asia*. Vol. 2, *Economic and Social Change*. Kuala Lumpur: Oxford University Press, 1979.

Teclaff, Ludwik A. "Harmonizing Water Use and Development with Environmental Protection." In *Water in a Developing World*, ed. Albert A. Utton and Ludwik Teclaff, pp. 72-125. Boulder: Westview Press, 1978.

_______. *The River Basin in History and Law*. The Hague: Martinus Nijhoff, 1967.

Teclaff, Ludwik A., and Teclaff, Eileen. "A History of Water Development and Environmental Quality." In *Environmental Quality and Water Development*, ed. Charles R. Goldman et al., pp. 26-77. San Francisco: W.H. Freeman and Co., 1973.

Thomas, L. William, Jr., et al., eds. *Man's Role in Changing the Face of the Earth*. Chicago: University of Chicago Press, 1955.

Tolentino, Amado S., Jr. "Philippine Coastal Zone Management: Organizational Linkages and Interconnections." *Environmental and Policy Institute Working Paper*. Honolulu: East-West Center, 1983.

Torry, William I. "Hazards, Hazes and Holes: A Critique of the Environment as Hazard and General Reflections on Disaster Research." *Canadian Geographer* 23 (Winter 1979): pp. 368-383.

United Nations. *Report of the United Nations Conference on the Human Environment*. New York: United Nations, 1973.

_______. Economic Commission for Europe. *Seminar on River Basin Management*. New York: United Nations, 1971.

United Nations Educational, Scientific and Cultural Organization. *Task Force on the Contribution of the Social Sciences to the MAB Programme: Final Report*. Paris: UNESCO, 1974.

University of San Carlos Water Resources Center. *Hydrology in Review Region VII*. Cebu City: Water Resources Center, 1984.

Valencia, Finina N. "Survey of the Status of Small Scale Backyard Livestock Raising in Siquijor." B.A. thesis, Foundation University, 1979.

Waddell, Eric. "The Hazards of Scientism: A Review Article." *Human Ecology* 5 (March 1977): 69-76.

Wernstedt, Frederick L., and Spencer, J.E. *The Philippine Island World.* Berkeley: University of California Press, 1967.

Wescoat, James L., Jr., "The 'Practical Range of Choice' in Water Resources Geography," *Progress in Human Geography* 11 (March 1987): 41-59.

White, Gilbert F. "Comparative Analysis of Complex River Development." In *Environmental Effects of Complex River Development,* ed. Gilbert F. White, pp. 1-21. Boulder: Westview Press, 1977.

_______. "Contributions of Geographical Analysis to River Basin Development." In *Readings in Resource Management and Conservation,* ed. Ian Burton and Robert W. Kates, pp. 375-394. Chicago: University of Chicago Press, 1963.

_______. "Natural Hazards Research." In *Directions in Geography,* ed. R.J. Chorley, pp. 193-216. London: Methuen and Co., 1973.

_______. "Role of Geography in Water Resources Management." In *Man and Water,* ed. Douglas James, pp. 102-121. Lexington: University of Kentucky Press, 1974.

Whyte, Anne. "The Integration of Natural and Social Sciences in the MAB Programme." *International Social Science Journal* 34 (1982): 411-426.

Worcester, Dean C. *The Philippine Islands and Their People.* London: Macmillan, 1899.

_______. "Contributions to Philippine Ornithology." *Proceedings of the U.S. National Museum* 20 (1898): 567-615.

World Bank. *Aspects of Poverty in the Philippines: A Review and Assessment.* Vol. 2. Washington, D.C.: World Bank, 1980.

Zelinsky, Wibur; Kosinski, Leszek A.; and Prothero, R. Mansell, eds. *Geography and a Crowding World.* New York: Oxford University Press, 1970.

Official Publications of the Republic of the Philippines

Department of Agriculture and Food. Bureau of Soils and Water Management. Agricultural Lands Management Evaluation Division. *Slope and Elevation Maps. Province of Siquijor. 1:50,000.* Manila: Bureau of Soils and Water Management, 1985.

_______. *Soil/Land Resources Evaluation for Agriculture: Province of Siquijor, Region 7.* Manila: Bureau of Soils, 1985.

Department of Agriculture and Natural Resources. Bureau of Soils. *Soil Survey of Negros Oriental,* by Alfredo Barrera and Jose Jaug. Soil Report no. 26. Manila: Bureau of Printing, 1960.

Department of Commerce and Industry. Bureau of the Census and Statistics. *Census of the Philippines. Agriculture 1960.* Vol. 1, *Agriculture Report by Province.* Manila: Bureau of Printing, 1960.

_______. *Census of the Philippines: 1948.* Vol. 2, pt. 3, *Report by Province for Census of Agriculture.* Manila: Bureau of Printing, 1953.

_______. *Net Internal Migration in the Philippines, 1960-1970,* by Yun Kim. Technical Paper no. 2. Manila: Bureau of the Census and Statistics, 1972.

_______. Bureau of Mines. *Preliminary Report of the Geology and Manganese Deposits of Siquijor Island, Negros Oriental, Philippines,* by Ronald K. Sorem. Manila: Bureau of Mines, 1951.

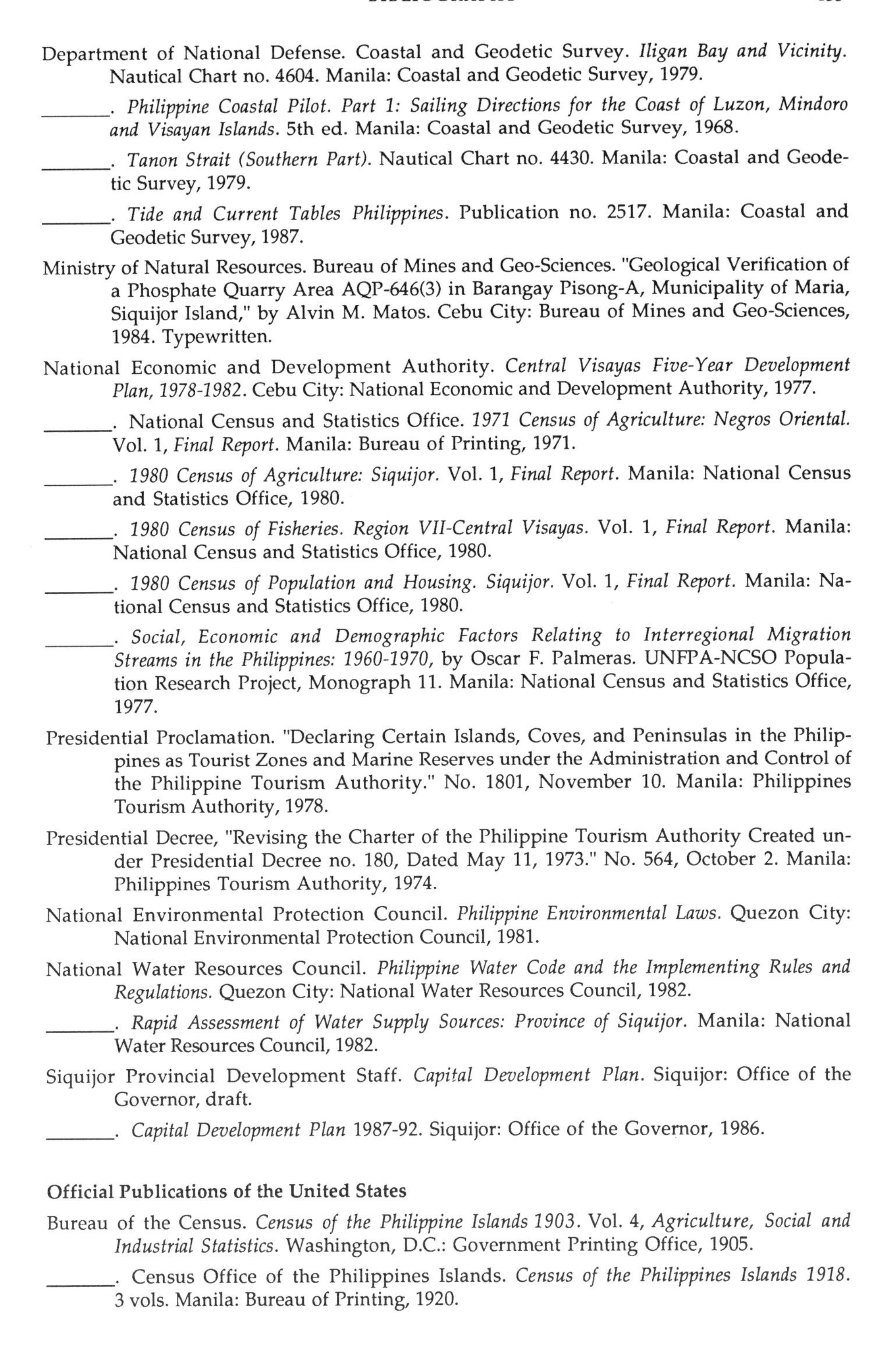

Department of National Defense. Coastal and Geodetic Survey. *Iligan Bay and Vicinity.* Nautical Chart no. 4604. Manila: Coastal and Geodetic Survey, 1979.

_______. *Philippine Coastal Pilot. Part 1: Sailing Directions for the Coast of Luzon, Mindoro and Visayan Islands.* 5th ed. Manila: Coastal and Geodetic Survey, 1968.

_______. *Tanon Strait (Southern Part).* Nautical Chart no. 4430. Manila: Coastal and Geodetic Survey, 1979.

_______. *Tide and Current Tables Philippines.* Publication no. 2517. Manila: Coastal and Geodetic Survey, 1987.

Ministry of Natural Resources. Bureau of Mines and Geo-Sciences. "Geological Verification of a Phosphate Quarry Area AQP-646(3) in Barangay Pisong-A, Municipality of Maria, Siquijor Island," by Alvin M. Matos. Cebu City: Bureau of Mines and Geo-Sciences, 1984. Typewritten.

National Economic and Development Authority. *Central Visayas Five-Year Development Plan, 1978-1982.* Cebu City: National Economic and Development Authority, 1977.

_______. National Census and Statistics Office. *1971 Census of Agriculture: Negros Oriental.* Vol. 1, *Final Report.* Manila: Bureau of Printing, 1971.

_______. *1980 Census of Agriculture: Siquijor.* Vol. 1, *Final Report.* Manila: National Census and Statistics Office, 1980.

_______. *1980 Census of Fisheries. Region VII-Central Visayas.* Vol. 1, *Final Report.* Manila: National Census and Statistics Office, 1980.

_______. *1980 Census of Population and Housing. Siquijor.* Vol. 1, *Final Report.* Manila: National Census and Statistics Office, 1980.

_______. *Social, Economic and Demographic Factors Relating to Interregional Migration Streams in the Philippines: 1960-1970,* by Oscar F. Palmeras. UNFPA-NCSO Population Research Project, Monograph 11. Manila: National Census and Statistics Office, 1977.

Presidential Proclamation. "Declaring Certain Islands, Coves, and Peninsulas in the Philippines as Tourist Zones and Marine Reserves under the Administration and Control of the Philippine Tourism Authority." No. 1801, November 10. Manila: Philippines Tourism Authority, 1978.

Presidential Decree, "Revising the Charter of the Philippine Tourism Authority Created under Presidential Decree no. 180, Dated May 11, 1973." No. 564, October 2. Manila: Philippines Tourism Authority, 1974.

National Environmental Protection Council. *Philippine Environmental Laws.* Quezon City: National Environmental Protection Council, 1981.

National Water Resources Council. *Philippine Water Code and the Implementing Rules and Regulations.* Quezon City: National Water Resources Council, 1982.

_______. *Rapid Assessment of Water Supply Sources: Province of Siquijor.* Manila: National Water Resources Council, 1982.

Siquijor Provincial Development Staff. *Capital Development Plan.* Siquijor: Office of the Governor, draft.

_______. *Capital Development Plan* 1987-92. Siquijor: Office of the Governor, 1986.

Official Publications of the United States

Bureau of the Census. *Census of the Philippine Islands 1903.* Vol. 4, *Agriculture, Social and Industrial Statistics.* Washington, D.C.: Government Printing Office, 1905.

_______. Census Office of the Philippines Islands. *Census of the Philippines Islands 1918.* 3 vols. Manila: Bureau of Printing, 1920.

Defense Mapping Agency. *Larena, Siquijor Island.* Topographic Map Sheet no. 3647 I. U.S. Army Map Series 733. Washington, D.C.: Defense Mapping Agency, 1956.

_______. *Lazi, Siquijor Island.* Topographic Map Sheet no. 3747 III. U.S. Army Map Series 733. Washington, D.C.: Defense Mapping Agency, 1956.

_______. *Maria, Siquijor Island.* Topographic Map Sheet no. 3747 IV. U.S. Army Map Series 733. Washington, D.C.: Defense Mapping Agency, 1956.

Geological Survey. *Reconnaissance Study of Sediment Transported by Stream Island of Oahu,* by B.L. Jones, R.H. Nakahara, and S.S.W. Chinn. Circular C33. Oahu: U.S. Geological Survey, 1971.

War Department. *Eighth Annual Report of the Philippine Commission,* part 1. Washington, D.C.: Government Printing Office, 1907.

_______. *Report of the Philippine Commission,* part 1, *1908.* Washington, D.C.: Government Printing Office, 1909.

_______. Bureau of Insular Affairs. *A Pronouncing Gazetteer and Geographical Dictionary of the Philippine Islands.* Washington, D.C.: Government Printing Office, 1902.

_______. *Report of the Bureau of Forestry of the Philippines Islands, Appendix J., Report of the Chief of the Forestry Bureau for the Period from July 1, 1901, to September 1, 1902.* Washington, D.C.: Government Printing Office, 1902.

_______. *Reports of the Philippine Commission.* Washington, D.C.: Government Printing Office, 1904.

Index

THE UNIVERSITY OF CHICAGO

GEOGRAPHY RESEARCH PAPERS

(Lithographed, 6 x 9 inches)

Titles in Print

127. GOHEEN, PETER G. *Victorian Toronto, 1850 to 1900: Pattern and Process of Growth.* 1970. xiii + 278 p.
130. GLADFELTER, BRUCE G. *Meseta and Campina Landforms in Central Spain: A Geomorphology of the Alto Henares Basin.* 1971. xii + 204 p.
131. NEILS, ELAINE M. *Reservation to City: Indian Migration and Federal Relocation.* 1971. x + 198 p.
132. MOLINE, NORMAN T. *Mobility and the Small Town, 1900-1930.* 1971. ix + 169 p.
133. SCHWIND, PAUL J. *Migration and Regional Development in the United States, 1950-1960.* 1971. x + 170 p.
134. PYLE, GERALD F. *Heart Disease, Cancer and Stroke in Chicago: A Geographical Analysis with Facilities, Plans for 1980.* 1971. ix + 292 p.
136. BUTZER, KARL W. *Recent History of an Ethiopian Delta: The Omo River and the Level of Lake Rudolf.* 1971. xvi + 184 p.
139. McMANIS, DOUGLAS R. *European Impressions of the New England Coast, 1497-1620.* 1972. viii + 147 p.
140. COHEN, YEHOSHUA S. *Diffusion of an Innovation in an Urban System: The Spread of Planned Regional Shopping Centers in the United States, 1949-1968.* 1972. ix + 136 p.
141. MITCHELL, NORA. *The Indian Hill-Station: Kodaikanal.* 1972. xii + 199 p.
142. PLATT, RUTHERFORD H. *The Open Space Decision Process: Spatial Allocation of Costs and Benefits.* 1972. xi + 189 p.
143. GOLANT, STEPHEN M. *The Residential Location and Spatial Behavior of the Elderly: A Canadian Example.* 1972. xv + 226 p.
144. PANNELL, CLIFTON W. *T'ai-Chung, T'ai-wan: Structure and Function.* 1973. xii + 200 p.
145. LANKFORD, PHILIP M. *Regional Incomes in the United States, 1929-1967: Level, Distribution, Stability, and Growth.* 1972. x + 137 p.
146. FREEMAN, DONALD B. *International Trade, Migration, and Capital Flows: A Quantitative Analysis of Spatial Economic Interaction.* 1973. xiv + 201 p.
147. MYERS, SARAH K. *Language Shift among Migrants to Lima, Peru.* 1973. xiii + 203 p.
148. JOHNSON, DOUGLAS L. *Jabal al-Akhdar, Cyrenaica: An Historical Geography of Settlement and Livelihood.* 1973. xii + 240 p.
149. YEUNG, YUE-MAN. *National Development Policy and Urban Transformation in Singapore: A Study of Public Housing and the Marketing System.* 1973. x + 204 p.
150. HALL, FRED L. *Location Criteria for High Schools: Student Transportation and Racial Integration.* 1973. xii + 156 p.
151. ROSENBERG, TERRY J. *Residence, Employment, and Mobility of Puerto Ricans in New York City.* 1974. xi + 230 p.
152. MIKESELL, MARVIN W., ed. *Geographers Abroad: Essays on the Problems and Prospects of Research in Foreign Areas.* 1973. ix + 296 p.
153. OSBORN, JAMES. *Area, Development Policy, and the Middle City in Malaysia.* 1974. x + 291 p.

154. WACHT, WALTER F. *The Domestic Air Transportation Network of the United States.* 1974. ix + 98 p.
155. BERRY, BRIAN J. L. et al. *Land Use, Urban Form and Environmental Quality.* 1974. xxiii + 440 p.
156. MITCHELL, JAMES K. *Community Response to Coastal Erosion: Individual and Collective Adjustments to Hazard on the Atlantic Shore.* 1974. xii + 209 p.
157. COOK, GILLIAN P. *Spatial Dynamics of Business Growth in the Witwatersrand.* 1975. x + 144 p.
160. MEYER, JUDITH W. *Diffusion of an American Montessori Education.* 1975. xi + 97 p.
162. LAMB, RICHARD F. *Metropolitan Impacts on Rural America.* 1975. xii + 196 p.
163. FEDOR, THOMAS STANLEY. *Patterns of Urban Growth in the Russian Empire during the Nineteenth Century.* 1975. xxv + 245 p.
164. HARRIS, CHAUNCY D. *Guide to Geographical Bibliographies and Reference Works in Russian or on the Soviet Union.* 1975. xviii + 478 p.
165. JONES, DONALD W. *Migration and Urban Unemployment in Dualistic Economic Development.* 1975. x + 174 p.
166. BEDNARZ, ROBERT S. *The Effect of Air Pollution on Property Value in Chicago.* 1975. viii + 111 p.
167. HANNEMANN, MANFRED. *The Diffusion of the Reformation in Southwestern Germany, 1518-1534.* 1975. ix + 235 p.
168. SUBLETT, MICHAEL D. *Farmers on the Road: Interfarm Migration and the Farming of Noncontiguous Lands in Three Midwestern Townships. 1939-1969.* 1975. xiii + 214 p.
169. STETZER, DONALD FOSTER. *Special Districts in Cook County: Toward a Geography of Local Government.* 1975. xi + 177 p.
171. SPODEK, HOWARD. *Urban-Rural Integration in Regional Development: A Case Study of Saurashtra, India—1800-1960.* 1976. xi + 144 p.
172. COHEN, YEHOSHUA S., and BRIAN J. L. BERRY. *Spatial Components of Manufacturing Change.* 1975. vi + 262 p.
173. HAYES, CHARLES R. *The Dispersed City: The Case of Piedmont, North Carolina.* 1976. ix + 157 p.
174. CARGO, DOUGLAS B. *Solid Wastes: Factors Influencing Generation Rates.* 1977. 100 p.
176. MORGAN, DAVID J. *Patterns of Population Distribution: A Residential Preference Model and Its Dynamic.* 1978. xiii + 200 p.
177. STOKES, HOUSTON H.; DONALD W. JONES; and HUGH M. NEUBURGER. *Unemployment and Adjustment in the Labor Market: A Comparison between the Regional and National Responses.* 1975. ix + 125 p.
180. CARR, CLAUDIA J. *Pastoralism in Crisis. The Dasanetch and Their Ethiopian Lands.* 1977. xx + 319 p.
181. GOODWIN, GARY C. *Cherokees in Transition: A Study of Changing Culture and Environment Prior to 1775.* 1977. ix + 207 p.
183. HAIGH, MARTIN J. *The Evolution of Slopes on Artificial Landforms, Blaenavon, U.K.* 1978. xiv + 293 p.
184. FINK, L. DEE. *Listening to the Learner: An Exploratory Study of Personal Meaning in College Geography Courses.* 1977. ix + 186 p.
185. HELGREN, DAVID M. *Rivers of Diamonds: An Alluvial History of the Lower Vaal Basin, South Africa.* 1979. xix + 389 p.

186. BUTZER, KARL W., ed. *Dimensions of Human Geography: Essays on Some Familiar and Neglected Themes.* 1978. vii + 190 p.
187. MITSUHASHI, SETSUKO. *Japanese Commodity Flows.* 1978. x + 172 p.
188. CARIS, SUSAN L. *Community Attitudes toward Pollution.* 1978. xii + 211 p.
189. REES, PHILIP M. *Residential Patterns in American Cities: 1960.* 1979. xvi + 405 p.
190. KANNE, EDWARD A. *Fresh Food for Nicosia.* 1979. x + 106 p.
192. KIRCHNER, JOHN A. *Sugar and Seasonal Labor Migration: The Case of Tucumán, Argentina.* 1980. xii + 174 p.
194. HARRIS, CHAUNCY D. *Annotated World List of Selected Current Geographical Serials, Fourth Edition. 1980.* 1980. iv + 165 p.
196. LEUNG, CHI-KEUNG, and NORTON S. GINSBURG, eds. *China: Urbanizations and National Development.* 1980. ix + 283 p.
197. DAICHES, SOL. *People in Distress: A Geographical Perspective on Psychological Well-being.* 1981. xiv + 199 p.
198. JOHNSON, JOSEPH T. *Location and Trade Theory: Industrial Location, Comparative Advantage, and the Geographic Pattern of Production in the United States.* 1981. xi + 107 p.

199-200. STEVENSON, ARTHUR J. *The New York–Newark Air Freight System.* 1982. xvi + 440 p.

201. LICATE, JACK A. *Creation of a Mexican Landscape: Territorial Organization and Settlement in the Eastern Puebla Basin, 1520-1605.* 1981. x + 143 p.
202. RUDZITIS, GUNDARS. *Residential Location Determinants of the Older Population.* 1982. x + 117 p.
203. LIANG, ERNEST P. *China: Railways and Agricultural Development, 1875-1935.* 1982. xi + 186 p.
204. DAHMANN, DONALD C. *Locals and Cosmopolitans: Patterns of Spatial Mobility during the Transition from Youth to Early Adulthood.* 1982. xiii + 146 p.
206. HARRIS, CHAUNCY D. *Bibliography of Geography. Part II: Regional. Volume 1. The United States of America.* 1984. viii + 178 p.

.207-208. WHEATLEY, PAUL. *Nagara and Commandery: Origins of the Southeast Asian Urban Traditions.* 1983. xv + 472 p.

209. SAARINEN, THOMAS F.; DAVID SEAMON; and JAMES L. SELL, eds. *Environmental Perception and Behavior: An Inventory and Prospect.* 1984. x + 263 p.
210. WESCOAT, JAMES L., JR. *Integrated Water Development: Water Use and Conservation Practice in Western Colorado.* 1984. xi + 239 p.
211. DEMKO, GEORGE J., and ROLAND J. FUCHS, eds. *Geographical Studies on the Soviet Union: Essays in Honor of Chauncy D. Harris.* 1984. vii + 294 p.
212. HOLMES, ROLAND C. *Irrigation in Southern Peru: The Chili Basin.* 1986. ix + 199 p.
213. EDMONDS, RICHARD LOUIS. *Northern Frontiers of Qing China and Tokugawa Japan: A Comparative Study of Frontier Policy.* 1985. xi + 209 p.
214. FREEMAN, DONALD B., and GLEN B. NORCLIFFE. *Rural Enterprise in Kenya: Development and Spatial Organization of the Nonfarm Sector.* 1985. xiv + 180 p.
215. COHEN, YEHOSHUA S., and AMNON SHINAR. *Neighborhoods and Friendship Networks: A Study of Three Residential Neighborhoods in Jerusalem.* 1985. ix + 137 p.
216. OBERMEYER, NANCY J. *Bureaucrats, Clients, and Geography: The Bailly Nuclear Power Plant Battle in Northern Indiana.* 1989. x + 135 p.

217-218. CONZEN, MICHAEL P., ed. *World Patterns of Modern Urban Change: Essays in Honor of Chauncy D. Harris*. 1986. x + 479 p.

219. KOMOGUCHI, YOSHIMI. *Agricultural Systems in the Tamil Nadu: A Case Study of Peruvalanallur Village*. 1986. xvi + 175 p.

220. GINSBURG, NORTON; JAMES OSBORN; and GRANT BLANK. *Geographic Perspectives on the Wealth of Nations*. 1986. ix + 133 p.

221. BAYLSON, JOSHUA C. *Territorial Allocation by Imperial Rivalry: The Human Legacy in the Near East*. 1987. xi + 138 p.

222. DORN, MARILYN APRIL. *The Administrative Partitioning of Costa Rica: Politics and Planners in the 1970s*. 1989. xi + 126 p.

223. ASTROTH, JOSEPH H., JR. *Understanding Peasant Agriculture: An Integrated Land-Use Model for the Punjab*. 1990. xiii + 173 p.

224. PLATT, RUTHERFORD H.; SHEILA G. PELCZARSKI; and BARBARA K. BURBANK, eds. *Cities on the Beach: Management Issues of Developed Coastal Barriers*. 1987. vii + 324 p.

225. LATZ, GIL. *Agricultural Development in Japan: The Land Improvement District in Concept and Practice*. 1989. viii + 135 p.

226. GRITZNER, JEFFREY A. *The West African Sahel: Human Agency and Environmental Change*. 1988. xii + 170 p.

227. MURPHY, ALEXANDER B. *The Regional Dynamics of Language Differentiation in Belgium: A Study in Cultural-Political Geography*. 1988. xiii + 249 p.

228-229. BISHOP, BARRY C. *Karnali under Stress: Livelihood Strategies and Seasonal Rhythms in a Changing Nepal Himalaya*. 1990. xviii + 460 p.

230. MUELLER-WILLE, CHRISTOPHER. *Natural Landscape Amenities and Suburban Growth: Metropolitan Chicago, 1970-1980*. 1990. xi + 153 p.

231. WILKINSON, M. JUSTIN. *Paleoenvironments in the Namib Desert: The Lower Tumas Basin in the Late Cenozoic*. 1990.

232. DUBOIS, RANDOM. *Soil Erosion in a Coastal River Basin: A Case Study from the Philippines*. 1990. xii + 138 p.